RESEARCH METHODS IN DIGITAL FOOD STUDIES

This book offers the first methodological synthesis of digital food studies. It brings together contributions from leading scholars in food and media studies and explores research methods from textual analysis to digital ethnography and action research.

In recent times, digital media has transformed our relationship with food which has become one of the central topics in digital and social media. This spatiotemporal shift in food cultures has led us to reimagine how we engage in different practices related to food as consumers. The book examines the opportunities and challenges that the new digital era of food studies presents and what methodologies are employed to study the changed dynamics in this field. These methodologies provide insights into how restaurant reviews, celebrity webpages, the blogosphere and YouTube are explored, as well as how to analyse digital archives, digital soundscapes and digital food activism and a series of approaches to digital ethnography in food studies. The book presents straightforward ideas and suggestions for how to get started on one's own research in the field through well-structured chapters that include several pedagogical features.

Written in an accessible style, the book will serve as a vital point of reference for both experienced researchers and beginners in the digital food studies field, health studies, leisure studies, anthropology, sociology, food sciences, and media and communication studies.

Jonatan Leer is head of food and tourism research at the University College Absalon and has published widely on food culture including *Food and Age* and *Alternative Food Politics*, and previously edited the anthology *Food and Media*. He is visiting lecturer at the University of Gastronomic Sciences, Pollenzo, Italy.

Stinne Gunder Strøm Krogager is Associate Professor at the Department of Communication and Psychology, Aalborg University, Denmark. She has published on food, gender and methodologies in, for example, Routledge's *Critical Food Studies*, and she is also Editor-in-Chief at the Nordic Journal, *MedieKultur: Journal of Media and Communication Research*.

RESEARCH METHODS IN DIGITAL FOOD STUDIES

Edited by Jonatan Leer and Stinne Gunder Strøm Krogager

Routledge
Taylor & Francis Group
LONDON AND NEW YORK

First published 2021
by Routledge
2 Park Square, Milton Park, Abingdon, Oxon OX14 4RN

and by Routledge
52 Vanderbilt Avenue, New York, NY 10017

Routledge is an imprint of the Taylor & Francis Group, an informa business

British Library Cataloguing-in-Publication Data
A catalogue record for this book is available from the British Library

Library of Congress Cataloging-in-Publication Data
A catalog record has been requested for this book

ISBN: 978-0-367-81926-2 (hbk)
ISBN: 978-0-367-81927-9 (pbk)
ISBN: 978-1-003-01084-5 (ebk)

Typeset in Bembo
by SPi Global, India

We would like to dedicate this volume to Dr. Karen Klitgaard Povlsen who introduced us both to food studies and brought us together. Her mentorship and her long engagement in the study of food and media have been a great source of inspiration for us and one of the main reasons for us undertaking the task of making this volume. Thank you!

Stinne and Jonatan

CONTENTS

FIGURES

TABLES

CONTRIBUTORS

Kerry Chamberlain (K.Chamberlain@massey.ac.nz) is Professor Emeritus (Health and Social Psychology) at Massey University Auckland, and Adjunct Professor (Sociology) at Victoria University of Wellington, New Zealand. He has research interests in media, health, and medicine in everyday life, food and health, and innovative qualitative research methods.

Alice Chik (alice.chik@mq.edu.au) is Senior Lecturer in the School of Education and Associate Director of Multilingualism Research Centre, Macquarie University, Australia. Her research interests include language learning in informal and digital contexts and multilingualism as urban diversities.

Karin Eli (karin.eli@anthro.ox.ac.uk) is a medical anthropologist and senior research fellow at Warwick Medical School, University of Warwick, UK. She is also deputy director of the Unit for Biocultural Variation and Obesity at the University of Oxford's School of Anthropology and Museum Ethnography.

Nicolai Jørgensgaard Graakjær (nicolaig@hum.aau.dk) is Professor of Music and Sound in Market Communication in the Department of Communication and Psychology at Aalborg University, Denmark. His research interests span from musicology, sound studies, and media studies to social psychology.

Rachael Kent (rachael.c.kent@kcl.ac.uk) is a Lecturer in Digital Economy & Society Education (Department of Digital Humanities), at King's College London, UK, and a Digital Health Consultant in the private sector. Kent's first book, *The Health Self: Digital Performativity and Health Management in Everyday Life* will be published by Bristol University Press in 2021. Kent's interdisciplinary research explores the relationship between digital media and communication, focusing specifically upon the intersections of technology and the body, health, and surveillance.

Katrine Meldgaard Kjær (kakj@itu.dk) is an assistant professor at the IT University of Copenhagen, Denmark. She has published work on the relationship between food and celebrity, and is in her current work interested in interdisciplinary research using computational and digital methods, particularly how qualitative methods may inform quantitative methods here.

Stinne Gunder Strøm Krogager (stinne@hum.aau.dk) is Associate Professor at the Department of Communication and Psychology, Aalborg University, Denmark. She has published on food, gender, and methodologies in, for example, *Critical Food Studies* (Routledge) and also, she is Editor-in-Chief at the Nordic Journal, *MedieKultur: Journal of Media and Communication Research.*

Jonatan Leer (jlee@pha.dk) is head of food and tourism research at the University College Absalon, Denmark, and has published widely on food culture including *Food and Age* and *Alternative Food Politics*, and previously edited the anthology *Food and Media*. He is visiting lecturer at the University of Gastronomic Sciences, Pollenzo, Italy.

Tania Lewis (tania.lewis@rmit.edu.au) is the Director of the Digital Ethnography Research Centre and Professor of Media and Communication at RMIT University, Australia. Her most recent book, *Digital Food: From Paddock to Platform* (Bloomsbury, 2020), is the first monograph to be published on everyday digital practices and food culture.

Deborah Lupton (d.lupton@unsw.edu.au) is SHARP Professor and leader of the Vitalities Lab, Centre for Social Research in Health and Social Policy Research Centre, University of New South Wales (UNSW) Sydney, Australia, and leader of the UNSW Node of the Australian Research Council Centre of Excellence for Automated Decision-Making and Society.

Meghan Lynch (meghan.lynch@mail.utoronto.ca) is a postdoctoral fellow at the Dalla Lana School of Public Health at the University of Toronto, Canada. Her research interests include exploring new methodologies and data sources, particularly through netnography, and on analysing policies and programs focused on improving the food environment.

Alana Mann (alana.mann@sydney.edu.au) is Associate Professor, Media and Communications, Faculty of Arts and Social Sciences (FASS), University of Sydney, Australia. Her latest book, *Food in a Changing Climate* (2021), challenges us to think beyond our plates to make our food systems more equitable and resilient.

Anders Kristian Munk (anderskm@hum.aau.dk) is Associate Professor at the Department of Culture and Learning at Aalborg University, Denmark, and Director of the Techno-Anthropology Lab. His work focuses on the development of digital and computational methods in the humanities with particular focus on controversy mapping.

Caroline Nyvang (cany@kb.dk) is a senior researcher with the Royal Danish Library. She received her Ph.D. on the History of Food in 2013 and has since published widely on how knowledge of food has been dispersed. Currently, Nyvang's research focuses on archived web.

Fabio Parasecoli (fabio.parasecoli@nyu.edu) is Professor of Food Studies in the Department of Nutrition and Food Studies at New York University, USA. His research explores food politics, media, and design. Recent books include *Knowing Food* (2019) and *Global Brooklyn: Designing Food Experiences in Global Cities* (2020).

Michelle Phillipov (michelle.phillipov@adelaide.edu.au) is a Senior Lecturer in Media at the University of Adelaide, Australia. Her research explores how media interest in food is shaping public debate, media and food industry practices, and consumer politics. She is the author or editor of four books.

Tanja Schneider (tanja.schneider@unisg.ch) is Associate Professor of Technology Studies at the School of Humanities and Social Sciences at the University of St. Gallen, Switzerland, and Research Affiliate at the Institute for Science, Innovation & Society (InSIS) at the University of Oxford, UK.

Thomas Mosebo Simonsen (tms@hum.aau.dk) is associate professor in the Department of Communication at Aalborg University, Denmark. He researches in communication and media sciences with a specific focus on digital media and film studies.

Camilla Vásquez (cvasquez@usf.edu) is a Professor of Applied Linguistics at the University of South Florida, USA. She is the author of *The Discourse of Online Consumer Reviews* (Bloomsbury, 2014) and her research on restaurant reviews has appeared in journals such as *Food & Foodways* and *Visual Communication*.

ACKNOWLEDGEMENTS

Jonatan Leer and **Stinne Gunder Strøm Krogager** would like to thank our editor Faye Leerink for her encouragement and dedication to this project. Also, we would like to thank Nonita Saha for all her hard work with finalising the manuscript. We would like to thank Janne Nielsen for her rapid feedback and help and Stine Mogenshøj Leer for her aesthetic guidance. Also, we would like to thank the three anonymous reviewers and Michael K. Goodman for their enthusiasm and their clever feedback, and Bente Halkier and Josée Johnston for their endorsements. We are also very grateful to all the contributors for both their brilliant work and professionalism; it has made this a wonderful and inspiring journey. Stinne would like to thank her great colleagues at the MÆRKK research group for their encouragement and constructive feedback. Jonatan would like to thank his amazing colleagues at BLM at the University College Absalon and those from Aarhus University who were a part of the action research project discussed in the book; Stig Brostrøm, Ole Henrik Hansen, Sarah Damgaard Warrer and Tine Mark Jensen, and all the dedicated practitioners in the project.

1

DOING DIGITAL FOOD STUDIES

Stinne Gunder Strøm Krogager and Jonatan Leer

Digital media has dramatically changed our relationship to food and food has become a predominant topic on digital and social media (Hu, Manikonda, & Kambhampati, 2014; Lewis, 2018). However, the literature on digital food studies is still relatively limited, despite innovative works appearing particularly since 2016 (maybe most importantly Leer & Povlsen, 2016; Lewis, 2020; Lupton & Feldman, 2020; Rousseau, 2012; Schneider, Eli, Dolan, & Ulijaszek, 2018). These works explore a range of issues such as how are new forms of food-political activism enabled on digital media? Or how are new digital practices integrated in our food practices from shopping to eating and evaluating food?

For instance, food photography and sharing images of what you eat have become an everyday activity – often mimicking the aesthetics of professionals and commercial food texts (Holmberg, Chaplin, Hillman, & Berg, 2016). Thus, the social act of eating has become closely related to the act of visually representing what you eat. The normalisation of sharing images in real time forces us to rethink the temporal and spatial boundaries of the meal. This additional ritual to the social act of eating together demonstrates how new digital technology not only facilitates the circulation of food representations but also shapes and transforms the way we eat and relate to food and meals.

Furthermore, food photography constitutes a form of productive consumption where the user becomes a "produser" (Bruns, 2008). The act of consumption is both personalised and distributed and the boundaries between producer and consumer are blurred. Also, aesthetic variations and taste serve to create social communities and mark boundaries to other communities following the Bourdieusian logic of food as a means to social distinction (Naccarato & Lebesco, 2012). It might be too simplified, however, to understand food photography merely as an act of social distinction. The digitalisation of food reaffirms affective relations and becomes a platform for affective expressivity (Lewis, 2018).

The digitalisation of food consumption is heavily debated. Is the digital era an age of liberation, global connectivity, and endless choices for the contemporary consumer? Or, rather, are the new digital tools just a sophisticated way of getting even closer to the dominance of big capitalism due to their digital surveillance and ability to manipulate the consumer in the most intimate manner? While this is a difficult argument to settle once and for all and, as both sides contain elements of truth, we must acknowledge that the digital age forces us to rethink food consumption and the role of the consumer.

As argued by Schneider et al. (2018), the digital era offers new forms of agency, connectivity, and activism across social and national boundaries that would have been far more difficult or time-consuming in the past. However, there are many kinds of digital food activism and these should always be considered in their political context as "diverse forms of digitally mediated practices of food activism" (p. 3). Also, the digital age offers new economic structures and new types of "alternative food politics" (Phillipov & Kirkwood, 2018) and these disrupt the traditional foodscapes and food systems through, for instance, alternative food networks and online shopping.

It seems fair to say that the literature on digital food studies has mainly focused on describing how the digitalisation of food practices have significantly transformed the mundane political and cultural aspects of food. We believe that the digital epoch of food studies also poses significant methodological challenges and potentials. In terms of potentials, the digital age offers new kinds of data sets (online debates, restaurants reviews, social media, etc.) as well as new kinds of tools to recruit informants and access material across the planet. In terms of challenges, the endless and ever-expanding platforms and amounts of new information can seem bewildering. These digital spheres are thus difficult to navigate with traditional research techniques and theories.

Also, how can we reconceptualise some very basic concepts previously taken for granted such as the concept of "text," which, in the context of social media, is no longer "fixed," but constantly evolving through new textual practices such as sharing, deleting, or commenting. This prompts fundamental epistemological discussions of hermeneutics and traditional semiotic models of communication as well as more instrumental issues of how to do research designs within the "digital." The "digital" also obliges researchers to revisit the ethical codex guiding their research and the way they interact with informants.

In this volume, we focus specifically on these methodological challenges and potentials and we draw on Jensen's understanding of "methodology" as a strategic level of analysis that connects practical matters with theoretical issues (Jensen, 2012), contrary to "methods" as concrete techniques for gathering and analysing data. Thus, the volume provides an overview of the research methods within digital food studies emphasising the concrete "how-to" dimension of doing digital food studies and supplementing this with methodological considerations. Hence, it appeals to both students and researchers in the social sciences and humanities as well as in the fields of nutrition and health and, of course, food studies, and gastronomy.

Digital food as a research interest and research area is relatively new and it arises from several different disciplines that draw on different methodological foundations. Consequently, the array of methods and tools being applied within digital food studies is diverse. However, we have identified four areas or approaches which characterise the field in general: textual analysis, digital ethnography, users' practices, digital archives, and network analysis. These concepts will guide the four chapters of this volume and we will explain them in more detail below.

Obviously, when doing digital food studies, we can draw on the wide range of books on methods within social science. However, they do not provide the knowledge specifically related to the field of digital food, let alone a straightforward and applicable how-to approach, which is exactly what this volume seeks to provide for both fellow researchers and students within the field of digital food studies.

Digital methods within the humanities and social sciences

As this volume will demonstrate, digital methods are plentiful and many are relatively interrelated. Various academic disciplines within the humanities and social sciences have turned to digital methods during the last couple of decades and this "digital turn" shows in publications such as Noortje Marres' *Digital Sociology* (Marres, 2017), Heather Horst and Daniel Miller's (eds.) *Digital Anthropology* (Horst & Miller, 2012) and Tom Boellstorff, Bonnie Nardi, Celia Pearce, and T. L. Taylor's *Ethnography and Virtual Worlds: A Handbook of Method* (Boellstorff, Nardi, Pearce, & Taylor, 2012). The digital turn within academic disciplines bears witness to the all-encompassing part played by "the digital" in modern life.

The digital turn within ethnographic research emerged around 2000 with the sociologist Christine Hine's *Virtual Ethnography* (Hine, 2000) and, following this, the interest in defining ethnography as "digital" has materialised primarily within the disciplines of anthropology and sociology (Pink et al., 2016). In *Digital Ethnography. Principles and Practice*, Sarah Pink, Heather Horst, John Postill, Larissa Hjort, Tania Lewis, and Jo Tacchi (2016) ask how the digital has redefined ethnographic practices. The authors apply a non-media-centric approach, which entails digital media and technologies being integrated parts of people's everyday lives inseparable from other everyday activities (Couldry, 2004; Hepp, 2012; Couldry & Hepp, 2013; Moores, 2018). Applying a non-media-centric approach thus implicates the understanding that (digital) media practices amalgamate (merge) with non-media practices in our everyday lives (Krogager, Leer, Povlsen, & Højlund, 2020), for instance, listening to podcasts while doing the dishes or looking up a recipe on the smartphone while cooking. Contrary to this, a media-centric approach would focus on the specific meanings and connotations of a particular media text. Thus, turning ethnography into digitalised everyday lives involves contact with participants often being mediated rather than based on direct presence; the researcher might track what people do online or be invited into the participants' social media practices; listening may include reading, and writing might be replaced by video, photography, or blogging (Pink et al., 2016: 3). Thus, applying a non-media-centric approach implies the stance that traditional media-centric research such as, for example, conventional textual analysis no longer apprehends the complexity of social life in digitalised societies.

A particular strand of digital ethnography is Internet ethnography, also defined as Netnography. In *Netnography: Redefined* (2015), the founder of the approach, Robert Kozinets, describes the phenomenon of netnographic studies as "the nature of online social experience and interaction" (Kozinets, 2015: 2), also, this description implies that netnography has been widely applied within the social science research traditions and the humanities. In their chapter in this volume, Lynch and Chamberlain illustrate how they have used netnography in a study of food blogs. They distinguish between passive and active netnography and apply a passive approach by simply observing discussions on food blogs to become immersed in, and gain an understanding of, the online posts of food bloggers. An active approach, on the other hand, would include using social media actively to involve participants in research.

Within the field of sociology, the seminal work of Sherry Turkle, *The Second Self: Computers and the Human Spirit* (Turkle, 2005), was one of the first works to introduce "the digital" as a sociological research interest. In this piece, Turkle sums up her own research on computer culture from the dawn of the personal computer in the 1970s, focusing on computers not just as tools, but as part of our everyday personal, social, and psychological lives. Researching the digital as a sociologist, Deborah Lupton has developed a four-fold typology that condenses the subdisciplines of digital sociology (Lupton, 2014), namely *Professional digital practice*: new digital tools as part of researchers' sociological practice; *Analyses of digital technology use*: researching how people use digital media, technologies, and tools, as well as the role of digital media in relation to institutions and social structures; *Digital data analyses:* using naturally occurring (qualitative or quantitative) digital data, and *Critical digital sociology:* engaging in reflexive analysis of the use of digital media (Lupton, 2014: 15–18).

During the past decade, digital anthropology has materialised as a subfield. It is important not to conflate anthropology with ethnography; however, most anthropologists will probably study the digital by means of an ethnographic approach (Pink et al., 2016: 6; see also Schneider and Eli's chapter in this volume). In the anthology *Digital anthropology*, Heather Horst and Daniel Miller propose six basic principles that define digital anthropology. The key standpoint of these principles combined is an opposition to all approaches that suggest that becoming digital has both rendered us less human, less authentic, and more mediated. Horst and Miller argue that we are just as human within the digital world and that the digital offers new prospects for anthropology to comprehend what it means to be human (2012: 3–4). This stance is also very much an underlying reason for embarking on the work of this volume. An approach to digital food offers new outlooks for scholarly work coming from all strands of the social sciences and humanities. Moreover, it offers potential synergies for merging different methodological approaches (see, for instance, the multimethod design in Schneider and Eli's chapter).

Around 2007, web-related media studies research was studied as more than just the web itself, namely as societal and cultural data, and this marks the computational turn or data studies turn (Rogers, 2019). Richard Rogers' *Doing Digital Methods* (2019) presents the state of the art in researching the natively digital and thus, how to undertake research by means of the web – not just about it. He does this through a rich set of case studies, for instance, *issue-crawling* that seeks to locate networks on the web using different link analysis software. Entering a list of URLs into the software, it "crawls" the URLs and finds links between them and also to other websites (p. 43). This method is applied in Anders Kristian Munk's chapter in this volume where Munk presents a case of how to build and analyse a network of Nordic chefs. Another example in Rogers' book is *website history* where the history of the web is approached using, for instance, the Internet archive. This resource and method is unfolded in Caroline Nyvang's chapter in the fourth part of this volume.

Digital methods and food studies

In food studies, the study of social and digital media has had a relatively slow start. The monograph *Food and Social Media* (Rousseau, 2012) raised some interesting question in relation to how the new media alter our approach to food and eating practices. Albeit pointing the finger at interesting dilemmas and transformation, Rousseau's book did not provide a deeper reflection on the methodological and scientific implication of this novel situation. De Solier's book *Food and the Self: Consumption, Production and Material Culture* (de Solier, 2013) offered a more stringent scientific reflection and an original ethnographic methodology in the investigation of how material culture and DIY culture *matter* to bloggers and food amateurs. De Solier's focus is on what people actually *did* with the media more than on the new media platforms themselves. She argues that the "value of adopting such an empirical approach to theory is that it gives us an understanding of what really matters in people's lives and what should, therefore, matter to social theory – as opposed to what we presume matters or what we think ought to matter" (pp. 164–165).

In the introduction to the 2016 anthology *Food and Media* (Leer & Povlsen, 2016), the authors call for an increased focus on media in food studies and propose a series of new theoretical tools to comprehend this "mediatisation" of food cultures. The chapters remain, however, mostly focused on the "old" media of print and television (with some exceptions such as Parasecoli, 2016; Povlsen, 2016).

After this rather slow and fragmented start, articles within digital food studies did not take off until the latter part of the 2010s, but have accelerated since 2017. As it appears from the overview in Lupton and Feldman (2020), the years 2017–2020 provide a critical mass of new, innovative studies (Lupton & Feldman, 2020). These include numerous articles and the important anthology *Digital Food Activism* (Schneider et al., 2018). This book calls for a stronger focus on how the digital era has also provided structural change to food systems and new potentials for collective activism. This flux of new research culminated in 2020 with two major publications:

Tania Lewis's monograph *Digital Food: From Paddock to Platform* and the anthology *Digital Food Cultures* edited by Deborah Lupton and Zeena Feldman.

Lewis's monograph covers a broad variety of topics and examples of digital food. The book investigates both structural changes in food practices due to digitalisation (such as new digitally structured forms of meal-sharing) as well as topics more related to identity politics (such as a chapter on the different forms of masculinity performed in the digital kitchens). Similarly, the anthology *Digital Food Cultures* covers a great variety of perspectives and case studies from distinct geographical contexts and media platforms. For instance, we find not only several articles on health communication and health communities online, but also contributions on how new culinary "amateur" influencers have transformed the restaurant business.

The contribution of this volume

The books mentioned above offer rich introductions to digital food studies and the many exciting scientific perspectives of the new digital world of food. Nonetheless, none of them address, at any length, the methodological implications of the digitalisation of food. It is precisely this gap in the methodological literature on digital food studies which the present volume seeks to cover.

We have asked people whom we find have contributed to the corpus of digital food studies to specifically examine their scientific *modus operandi* and reflect on their methodology. Notably, how did they work concretely and methodologically with digital food? What were the methodological challenges and possibilities? What were the ethical reflections before and after their empirical work? How did their work differ from and resemble other works within food studies?

Thus, this volume does not present much new empirical data, but a series of in-depth methodological reflections on how to do digital food studies. We believe that this is a very pressing issue in contemporary food studies due to the fact that the digitalisation of food affects all domains of food studies from ethnography of everyday practices to historical analyses of digitalised archives. The slow start of digital food studies up to 2016 might be explained by many researchers' methodological inquietude when faced with the overwhelming new digital foodscapes. This is quite understandable as the digital era raises many methodological questions such as the selection of digital samples, the ontological status of digital material, and the ethics of working with user-generated digital material. This volume provides many of the answers to these questions. The answers might differ and they might not be conclusive. However, we feel sure that, with this volume as a companion, you will travel into the world of digital food studies with a solid, methodological understanding.

The chapters of this volume

Coming, as they do, from different research areas and traditions within communication, media and cultural studies, history and techno-anthropology, the contributors frame different methodological challenges and thus also different potentials. However, the methodological approaches applied in the contributions are predominantly qualitative, which also mirror the field of food studies more generally, although lately, a focus on quantitative and mixed methods is gaining ground.

The contributions offer a broad survey of research methods within digital food studies. The survey is not exhaustive; still, it covers a wide range of different approaches to studying digital food. Most of the methods applied are "time-honoured" in the sense that we have, for instance, done textual analyses and ethnography for many years. However, applying these methods to ever-changing digital platforms, contexts, and texts poses new challenges and thus requires new ethical reflections and strategies in terms of preserving material. Other methods, for instance, network and hashtag analysis are new in the sense that they are linked to the digital era and thus have a shorter history.

Each chapter provides an insight into the core works in the field and the major developments of this subdiscipline. At the same time, the chapters give concrete examples of scientific approaches, research designs, and tools from the respective authors' own research in order to exemplify the specific "how to" dimension of doing digital food studies. Furthermore, each contribution ends with a number of pedagogical features summing up the key points and reflections of the chapter and providing readers with an option of trying out the method in question for themselves. We have asked all contributors to include ethical considerations as the digital also adds new dimensions to the ethical aspects of research as it does to social life in general.

Part 1: Textual analysis

Michelle Phillipov begins the first part by discussing how textual analysis in the digital era still functions as a highly productive method to comprehending the meanings of digital food media. Yet, comprehending the complexity of digital food texts requires us to look beyond the analysis of specific texts to reflect on how texts move across and between media platforms and how they relate to other texts and genres.

In the second chapter, Katrine Meldgaard Kjær argues that food has always been connected with lifestyle and identity. Also, the digitalisation of food and celebrities provides new potentials for those in the public eye to acquire status through food-related practices in the media. Hence, celebrities represent important voices in contemporary food culture, which should be taken seriously by food studies. Kjær offers a methodological toolbox rooted in post-structural theory to unpack celebrity-brand web shops as sites of meaning, identity, and power.

The third contribution demonstrates the importance of studying sounds when working with digital food. Taking a text-analytical stance informed by music semiotics and sound branding, Nicolai Jørgensgaard Graakjær accentuates the often overlooked aspect of sound in digital foods illustrated through a case promoting champagne.

In the last contribution, Thomas Mosebo Simonsen and Stinne Gunder Strøm Krogager discuss how to approach the major food platform YouTube from a text-analytical perspective. The authors cover a series of issues specific to the platform, notably how handling samples on YouTube is challenged by large amounts of data and how recommendatory systems and algorithms favour the cases towards popularity parameters. Therefore, Simonsen and Krogager explore how to link the conventional sampling with a qualitative contextual approach including social networks, users, and third-party platforms to recognise appropriate cases of food videos on YouTube.

Thus, across the different foci in each chapter, this section addresses how traditional textual and visual analysis can be revisited and productively applied in the field of digital food studies.

Part 2: Digital ethnography

Tanja Schneider and Karin Eli open this section with a contribution on how to conceptualise the digital in digital food studies. The authors present a review of recent studies on food and eating in the digital area with particular focus on their own and others' work on digital food activism. On this basis, they propose a multifaceted ethnographic method, mixing auto-ethnography, participant observation, semi-structured interviews, and media analysis to describe real-life experiences of digital food and digital eating.

In the second contribution, Alana Mann explores the potentials for hashtag ethnography as a research method. Hashtag activism occurs when big numbers of postings emerge on social media under a mutually hashtagged word, phrase, or sentence with a social or political statement. Mann depicts hashtag ethnography as a suitable method to investigate the consequences of digital practices in the field of food politics.

The third contribution deliberates on the possibilities and restraints of digital food activism. Tania Lewis uses participatory and digital ethnographic methods to show how food politics and the tools of the digital area have empowered numerous players to use digital platforms and social media to connect, organise, and increase the visibility of a variety of food issues.

Chapter 4 in this section explores food-blogging communities from a netnographic perspective. Meghan Lynch and Kerry Chamberlain describe blogs as a lens to comprehending how people are turning what was formerly personal information into public property. Lynch and Chamberlain discuss encounters in the field and outline practical skills and guidelines in, for instance, defining the suitability of examining food blogs and food blogging communities, ethical considerations, and ways to handle trustworthiness and quality of data.

Camilla Vásquez and Alice Chik complete the section by addressing the changes that digital media have brought to the restaurant review genre. Considering the numerous issues of research into restaurant reviews, the authors highlight developments in research methodologies and offer applied suggestions in relation to ethics, data collection, data management, and data analysis.

Hence, this section offers six different methodological approaches and empirical foci to doing digital ethnography within the field of digital food studies.

Part 3: Users' practices

Rachael Kent begins this section. She explores how self-tracking technologies and social media platforms facilitate a re-articulation, representation, and understanding of body and health. Kent suggests an innovative methodological approach to investigating online and offline negotiations by self-tracking users merging pre-digital ethnographic methods (semi-structured interviews and guided reflexive diaries) with digital ethnography (online data capture). This triangulation offers knowledge of how users mediate their lives through self-tracking and social media platforms, providing empirical research not yet undertaken within the field of digital food.

In the second contribution, Jonatan Leer discusses action research (AR) in digital food studies. AR is the most explicit form of change-making through research. The researcher takes part in local problems and critically reflects on how to do things better in a joint process with practitioners. This kind of research poses basic problems in the way that the researcher and the scientific objectives are framed and also as regards methodology and the status of data and data collection. However, Leer argues that digital devices can potentially solve several of the problems inherent in AR.

Fabio Parasecoli concludes this section by presenting a workshop and his research project Global Brooklyn: a recurring, loosely classified set of sensoria, practices, and discourses which occur in coffee shops and restaurants in cities all over the world. Parasecoli discusses how global digital media, and especially visual media, were essential in the growth of the research on Global Brooklyn, offering virtual access to faraway localities and thus, enabling the recruitment of authors for an anthology that studies Global Brooklyn worldwide. Implementing digital media in research processes offers research crowdsourcing not yet common in food studies.

Part 4: Digital archives and network analysis

Caroline Nyvang describes how to work with archived web pages as a resource in *Food History*. The Internet has been termed a medium that "never forgets" and that information from the Internet is fleeting, dynamic, and highly challenging to contain. This ephemeral nature of the Internet means that it needs to be archived before it can be studied. Centring on methodological potentials and trials, Nyvang studies how Internet archives can be successfully implemented in food studies.

Anders Kristian Munk outlines how we can study huge corpuses to discover trends in food culture. Through a corpus of 102 million posts from 242,000 public food pages, he analyses the New Nordic cuisine movement. The chapter demonstrates how digital patterns can be mapped, and uses the case as an occasion to discuss how traces from social media can tell us something about the societal impact of new food trends.

In the afterword, Deborah Lupton provides a more general perspective on Food Studies. Drawing on the different perspectives in the book and on her own recent anthology *Digital Food Cultures*, she addresses the methodological, theoretical, philosophical, and ethical issues that follow digitalisation.

Future encounters

The present volume demonstrates a great variety of methods in which to face the digitalisation of food culture and addresses the methodological implications. We would like to conclude by highlighting two aspects. Firstly, the digital era has changed many aspects of the way we talk, shop, and evaluate food and we need, as researchers, to take this very seriously – also in the way we operate methodologically. At the same time, it is important to note that not all has changed, and not all is new. A lot of the "good old" methods and tools still work. Central aspects of the good old techniques like textual analysis, observation, and interview are still valid; they just need to be reconsidered within the realm of the digital. A case in point is Phillipov's discussion of textual analysis in the opening chapter. She underlines, in relation to this discipline, that "while traditional approaches to textual analysis have maintained enduring value in a digital era, they also require some updating." This is true for most of the methods discussed in this volume, including ethnography, action research, political analysis, etc.

Secondly, this exploration gathers some of the most prominent researchers within this field today, but we do not pretend to present you with a conclusive list of all methods relevant to digital food studies. We cannot offer a final map of this undiscovered and rapidly changing field. Rather, we encourage readers to consider this as a series of informative and initial sketches of the new landscape. Something to inspire you on your journey into doing your own digital food study, but we also consider this volume an invitation to develop these methodological suggestions as you go along. There is much more to be explored in digital food studies and we will need many more innovative and creative methodologies for that exploration.

References

Boellstorff, T., Nardi, B., Pearce, C., & Taylor, T.L. (2012). *Ethnography and Virtual Worlds: A Handbook of Method*. New Jersey: Princeton University Press.

Bruns, A. (2008). *Blogs, Wikipedia, Second Life and Beyond: From Production to Produsage*. New York: Peter Lang Publishing.

Couldry, N. (2004). Theorising media as practice. *Social Semiotics*, 14(2): 115–132.

Couldry, N. & Hepp, A. (2013). Conceptualizing mediatization: Contexts, traditions, arguments. *Communication Theory*, 23(3): 191–202.

de Solier, I. (2013). *Food and the Self: Consumption, Production and Material Culture*. New York: Bloomsbury Publishing.

Hepp, A. (2012). Mediatization and the 'molding force' of the media. *Communications*, 37(1): 1–28.

Hine, C. (2000). *Virtual Ethnography*. London: Sage.

Holmberg, C., Chaplin, J., Hillman, T., & Berg, C. (2016). Adolescents' presentation of food in social media: An explorative study. *Appetite*, 99. January 2016. doi: 10.1016/j.appet.2016.01.009

Horst, H.A. & Miller, D. (Eds.). (2012). *Digital Anthropology*. Oxford: Berg Publishers.

Hu, Y., Manikonda, L., & Kambhampati, S. (2014). *What we instagram: A first analysis of instagram photo content and user types. Proceedings of the 8th International Conference on Weblogs and Social Media*, ICWSM 2014, pp. 595–598.

Jensen, K.B. (Ed.). (2012). *A Handbook of Media and Communication Research. Qualitative and Quantitative Methodologies*. London and New York: Routledge.

Kozinets, R.V. (2015). *Netnography: Redefined*. London: Sage.

Krogager, S.G.S., Leer, J., Povlsen, K.K., & Højlund, S. (2020). The amalgamation of media use practices and food practices in a school setting: Methodological reflections on doing non-media-centric media research with children. *Communication Research and Practice*, 6(2): 79–94.

Leer, J. & Povlsen, K.K. (Eds.). (2016). *Food and Media. Practices, Distinctions and Heterotopias*. London: Routledge.

Lewis, T. (2018). Digital food: From paddock to platform. *Communication Research and Practice*, 4(3): 212–228.

Lewis, T. (2020). *Digital Food: From Paddock to Platform*. London: Bloomsbury Academic.

Lupton, D. (2014). *Digital Sociology*. London and New York: Routledge.

Lupton, D. & Feldman, Z. (2020). *Digital Food Cultures*. London: Routledge.

Marres, N. (2017). *Digital Sociology*. Cambridge: Polity Press.

Moores, S. (2018). *Digital Orientations: Non-Media-Centric Media Studies and Non-Representational Theories of Practice*. New York: Peter Lang Publishing.

Naccarato, P. & Lebesco, K. (2012). *Culinary Capital*. New York: Berg Publishers.

Parasecoli, F. (2016). Manning the table: Masculinity and weight loss in U.S. commercials. In J. Leer & K.K. Povlsen (Eds.), *Food and Media. Practices, Distinctions and Heterotopias* (pp. 95–109). New York: Routledge.

Phillipov, M. & Kirkwood, K. (Eds.). (2018). *Alternative Food Politics: From the Margins to the Mainstream*. London: Routledge.

Pink, S., Horst, H., Postill, J., Hjorth, L., Lewis, T., & Tacchi, J. (2016). *Digital Ethnography. Principles and Practice*. London: Sage.

Povlsen, K.K. (2016). 'I (never) just google': Food and media practices. In J. Leer & K.K. Povlsen (Eds.), *Food and Media. Practices, Distinctions and Heterotopias* (pp. 129–148). New York: Routledge.

Rogers, R. (2019). *Doing Digital Methods*. London: Sage.

Rousseau, S. (2012). *Food and social media: You Are What You Tweet*. New York: Alta Mira Press.

Schneider, T., Eli, K., Dolan, C., & Ulijaszek, S. (Eds.). (2018). *Digital Food Activism*. London: Routledge.

Turkle, S. (2005). *The Second Self: Computers and the Human Spirit*. Cambridge: MIT Press.

PART 1

Textual analysis

2

TEXTUAL ANALYSIS IN DIGITAL FOOD STUDIES

New approaches to old methods

Michelle Phillipov

Studying digital food texts

The digital era has seen a proliferation in the number and types of media texts devoted to food. From food blogs, websites, and recipes to advertising, social media, and YouTube cooking shows, an array of food media texts is available at the click of a mouse or the tap of a screen. This volume and variety of digital food media texts makes it increasingly challenging for researchers to make sense of the complexity of digital food, including how digital food media "works" and how we might understand new and emerging relationships between media texts, messages, and platforms.

Although textual analysis has been much critiqued as a media and cultural studies research method for its inattention to the political economy of media production and to the nuances of audience interpretation (e.g., Philo, 2007), it remains a valuable tool for investigating what food, and food media, means in a digital age. The enduring value of textual methods is that they attune researchers to the ways in which representations invite and shape meaning, and to the power relationships involved in their construction and circulation. These questions of meaning and power are especially urgent in the contemporary context, given that digital food media is increasingly shaped by industry logics in which commercial and economic relationships are hidden from public view (Lewis, 2019; Phillipov, 2017).

This chapter provides a practical "how to" for conducting textual analysis in a digital era. It begins with a history of textual analysis, focusing on how "old" methods of textual interpretation are being adapted for a digital age. It then offers an overview of the main methods of textual analysis used in digital food studies, followed by practical examples and advice as to how these methods can be applied. There are many different ways of doing textual analysis, but as its core, textual analysis involves identifying the "most likely interpretations" that can be made of a text (McKee, 2003: 1). Any text can be subjected to textual analysis: blogs, advertising, television, social media, food product packaging, supermarket layouts, rural landscapes – the list goes on!

Central to the value of textual analysis for digital food studies is its flexibility and broad applicability. It allows researchers to identify repeated patterns across groups of texts, and to unpack the often "hidden" meanings of food texts. For example, textual analysis can reveal how the same types of idyllic rural imagery used to advance an alternative food politics reappear in the advertising strategies of major food manufacturers promoting industrial foods (e.g., Jackson et al., 2007). Likewise, it can show how the language of public health is appropriated by "Big Food" to present personal responsibility and (more) consumption as "solutions" to diet-related disease (Mayes, 2014). By closely examining word choices, imagery, narrative structures, signs, and symbols, textual analysis allows researchers to identify messages that audiences are invited to focus on and those that are obscured, thereby revealing both the "intertextual" (Frow, 2006; Kristeva, 1980) relationships between texts and the ways in which textual representations help to construct and circulate dominant ideas about food.

In doing so, textual analysis enables researchers to delve into meanings that may be beyond the conscious understanding of media producers and consumers. For example, textual analysis can show how racism, sexism, and speciesism appear, both unacknowledged and unrecognised, in texts ranging from vegan cookbooks and blogs (Harrington et al., 2019) to the social media marketing strategies of the Swedish dairy industry (Linné, 2016). The detail and insight that comes from close analysis of media texts, then, is one of the key benefits textual analysis offers to digital food studies.

Textual analysis: Then and now

Food studies initially applied textual analysis to individual food texts, such as cookbooks (e.g., Gallegos, 2005; Jones & Taylor, 2001), with analyses often focused on historically situating textual representations of particular foods, culinary taste cultures, cooking practices, and/or ethnic identities. During the 1990s and 2000s, the burgeoning literature on food television saw researchers apply textual analysis to television cooking shows to identify their genre and visual conventions (e.g., Chan, 2003; Strange, 1998), lifestyle discourses (e.g., Ketchum, 2005; Moseley, 2001), production of class distinctions (e.g., de Solier, 2005), and adherence to neoliberal logics (e.g., Lewis, 2008, 2011).

Other studies focused on celebrity chefs as "texts." Studies performed close readings of Jamie Oliver's performances of masculinity in *The Naked Chef* (Hollows, 2003b; Moseley, 2001), Hugh Fearnley-Whittingstall's class identity in *River Cottage* and *Hugh's Chicken Run* (Bell & Hollows, 2011), and Nigella Lawson's sexualised, "postfeminist" persona in *Nigella Bites* (Andrews, 2003; Hollows, 2003a). While there is still work continuing in this vein (e.g., Craig, 2019), single studies of individual texts now tend to appear less frequently in contemporary food studies.

The proliferation of media texts in a digital age necessitates new ways of doing textual analysis. Studies have moved away from deep engagement with individual texts to investigate a broad range of texts and their relationships. Tania Lewis's (2019) recent work, for instance, brings together analysis of food photography and video-sharing with discussion of online food activism and engagement by "Big Food" players to reveal the increasingly complex interdependencies and commercial logics that now inform and support everyday engagements with online food.

Since the 2000s, studies have paid greater attention to the ways in which texts move across and between different media genres and platforms. For example, celebrity chefs have been increasingly studied not as representations that exist "within" individual texts like television cooking shows, but as "entertainment packages" (Ashley et al., 2004: 175) that circulate across different media forms, including books, websites, social media, advertising, and product placements. For example, Jennifer Silver and Roberta Hawkins" (2017) study of sustainable seafood campaigns shows how celebrity personalities do not simply appear as the "face" of campaigns within individual media texts; rather, their actions shore up brand capital (for both the celebrity and the campaign) across a range of texts, including websites, magazines, and television shows, revealing the increasing centrality of celebrity to the enactment of contemporary environmental politics.

Methods for analysing media and digital texts

Perhaps more so than any other established method, textual analysis is especially varied in its methodological approach. Studies range from "anecdotal" forms of close reading (e.g., Chan, 2003) to much more "scientifically"-inclined content analyses (e.g., Vidal et al., 2015). The fact that the most influential food studies scholars using textual analysis tend to come from the humanities, rather than the social sciences, means that text-based work in food studies tends to have a less explicit discussion of method than is typical of other empirical approaches. Scholars using textual analysis tend to draw from the main textual methods used in media and cultural studies (for a taxonomy, see Weerakkody, 2009). Some of the most common – and most relevant to digital food studies – are outlined below.

Textual "readings"

Adapted from literary and cultural studies' methods of close reading, this is the textual method perhaps most commonly used in food studies. This approach to textual analysis tends not to follow formal methodological protocols, but instead offers analysis and interpretation of the significant details of, or patterns within, media and other texts. Because it is largely unrestricted by methodological rules that might narrow the researcher's field of interpretation, scholars undertaking such an openly "interpretive" approach can benefit from unexpected insights. For example, by paralleling depictions of "excessive" consumption in two seemingly different types of

YouTube videos – health influencer Stephanie Buttermore's "cheat day" videos and the gargantuan meals prepared on online cooking show *Epic Meal Time* – Deborah Lupton (2020) shows the "affective force" associated with transgressive eating and the ways in which such consumption can reveal cultural anxieties associated with eating the "wrong" foods.

Textual readings are anchored in theory and critique, which enables researchers to situate textual meanings in their wider political, social and industry contexts. In a series of studies, including a much-cited paper on *Jamie's Ministry of Food* (Hollows & Jones, 2010), Joanne Hollows and Steve Jones use textual analysis to trace the evolution of Jamie Oliver's public identity from "lifestyle expert" to "moral entrepreneur." Using his television cooking shows as examples, they show how Oliver's media work increasingly reframes food from a pleasurable leisure pursuit to something essential to the "moral redemption" of working class subjects, particularly a time of growing concern about obesity (Hollows & Jones, 2010: 309). In doing so, Hollows and Jones locate textual representations on television in relation to wider discourses of class pathologisation and the "obesity epidemic" also occurring in the UK at the time.

Discourse and framing analyses

In contrast to more open and interpretive textual "readings," discourse and framing analyses tend to be highly structured qualitative methods designed to identify deeper patterns, meanings, and power relations embedded in media texts. Discourse scholars analyse how language is used to enable some understandings of an issue while constraining others. Discourse analysts approach language as something that shapes and is shaped by society; that helps to constitute (and change) knowledge, social relations, and identities; and that is shaped by relations of power (Fairclough, 1995).

Discourses analyses of food television have investigated how discourses of personal and professional incompetency enable particular understandings of food waste on *Ramsay's Kitchen Nightmares* (Thompson & Haigh, 2017), and how "campaigning culinary documentaries" fronted by celebrity chefs employ discourses of responsibilisaton, governmentality, and neoliberalism to add moral authority to the chef's celebrity brands (Bell et al., 2017). Analyses of food blogs reveal how women bloggers employ discourses of food and femininity to construct and perform "saleable" identities in the blogosphere (Rodney et al., 2017).

Framing analysis investigates the way that media focuses attention on certain issues and places them within particular fields of meaning (see Entman, 1993). Scholars performing framing analyses identify the various "frames" or perpectives that structure media narratives. It is a method most commonly used for analysis of print and online news, with studies, for instance, examining how the regulation of fast food advertising (Henderson et al., 2009) or farm animal welfare (Buddle & Bray, 2019) are framed in print media, or how the "promises" of 3D printed food are framed in online news (Lupton, 2017). More broadly, framing analysis has been applied to the analysis of racial biases in online restaurant reviews (Zukin et al., 2017) and to the analysis of Twitter messages associated with the Meat Free Mondays campaign (Friedlander & Riedy, 2018).

Visual analysis

Visual analysis investigates how images communicate meaning. In digital food studies, researchers frequently employ methods of visual semiotics derived from Roland Barthes (1977) and other theories associated with the grammar of visual design (e.g., Kress & van Leeuwen, 1996). Scholars adopting this approach typically analyse how the composition and construction of images produce symbolic meanings, with studies ranging from analyses of digital "food porn" as post-feminist play (Dejmanee, 2016; see also the Introduction to this volume), to investigations of the use of monstrous "Frankenfood" memes in anti-GM campaigns (Clancy & Clancy, 2016), to analyses of how semiotic meanings are used in food packaging to "greenwash" food products (Wagner, 2015). While digital media scholars have been relatively slow in adopting visual methods (Highfield & Leaver, 2016), there has been a notable increase in digital food scholars analysing visual social media, particularly studies of food porn and digital images (e.g., Lupton, 2019; McDonnell, 2016).

Content analysis

Content analysis is used to identify and analyse the manifest content of media messages (Weerakkody, 2009). It is a systematic and quantitative approach that involves measuring the "amounts" of messages that appear in particular categories, for example, words, phrases, images, or themes. Content analysis is useful for analysing large numbers of media texts, as it enables the mapping of overall trends or changes in communication. As a result, it has enjoyed something of a renaissance in recent years with the rise of digital media and computer-assisted forms of analysis (Hansen & Machin, 2013). Content analysis is increasingly used for analysis of "big data" sets, particularly of social media, with traditional methods of content analysis supplemented with – and, in some cases, supplanted by – software-assisted algorithmic approaches (for a discussion, see Lewis et al., 2013).

Content analysis is widely used in digital food research, and has been especially dominant in health-related disciplines to examine, for example, how obesity is portrayed on YouTube (Yoo & Kim, 2012), the frequency of junk food marketing on Instagram (Vassallo et al., 2018), or the marketing techniques employed by food company websites (Hurwitz et al., 2017). Because it is primarily a quantitative method, content analysis allows researchers to identify *what* messages are present in a sample of texts, but not necessarily what these messages *mean*. Consequently, content analysis is often used in conjunction with other, more qualitative and interpretive, modes of analysis. For instance, Emily Contois (2017) combined content analysis with close reading in order to critique the images of hyperfeminine domesticity so frequently used to signify "health" in food blogs.

Multimethod approaches

As well as studies combining several methods of textual analysis, it is also now common for digital food studies to combine textual methods with other types of empirical research, such as interviews or observational analysis (for more detail on these approaches in digital food studies, see Lynch and Chamberlain, in this volume). For example, Michael Pennell's (2016) work on the use of Twitter by American food vendors and chefs combined interview and observational data about participants' Twitter activity with textual analysis of their Twitter posts to show the nuanced ways in which tweets and photos can contribute to the transparency of local and national food systems by "making visible" a range of food production processes. Combining textual and empirical methods in these ways, then, can substantially enrich a study's textual interpretations.

Research design for textual analysis: A practical example

To demonstrate in practical terms how textual analysis can be used, the following section outlines a step-by-step process for integrating textual methods into a large-scale research project. The approach I offer is a very personal one: the example comes from one of my own research projects (Phillipov, 2017), and outlines the "behind-the-scenes" thought processes that underpinned the research design and approach, what I looked for when conducting the textual analysis, and how the evidence I found helped me to think about my research questions. There are many ways that textual analysis can be successfully applied to food research, so this approach should not be taken as prescriptive. However, as an example, it usefully illustrates the various ways that food media texts produce meaning, and how these meanings are both constructed and altered as they move across media platforms and contexts.

The example is from a larger project investigating media's role in mainstreaming "alternative" food politics, as part of which I became interested in the way that the work of food production is represented in media. I suspected that, at a time of intensified media interest in food, we were also seeing more media interest in food production, and I wanted to better understand the implications of this for food workers.

To begin the study, I collected a range of Australian food-related texts over a 3-year period (2014–2016). The texts spanned a variety of mediums and platforms, including television cooking shows, news, advertising, cookbooks, and digital and social media. I began my close textual analysis with a single text: a cookbook called *It Tastes Better: Over 100 New Recipes Using My Favourite Sustainable Produce*. I started with this text because its discursive and visual tropes of "local," "seasonal," and "ethical" eating were typical of emerging food media trends in Australian and Anglophone media (for a more detailed discussion of method and text selection, see introductory chapter in Phillipov, 2017). This typicality was essential for how I justified a focus on a single text potentially unfamiliar to international audiences.

I first subjected the cookbook to discourse and visual analyses. Because my research question centred on representations of work, this was the focus of my interpretations. I considered: how were food workers represented? How was their work portrayed? Where was their work located? I noticed that, throughout the cookbook, farmers were typically presented in smiling, posed photos, while the action shots were reserved for images of chefs and restaurant staff. Analysing the written text, I noticed the highly affective language that farmers used to talk about their work: typically, it was not presented as "work" at all, but as a "passion" or a "labour of love." I collected and noted my examples.

Since this was not a project about the book itself, but rather one that used a specific text as a means to think through the relationships between the media representations and public discourses associated with food production work, I then moved outwards from this single "anchor" text to investigate the circulation of similar representational strategies in the wider media landscape.

I found, for example, almost identical imagery of smiling farmers and the language of "labours of love" in the advertising campaigns of major Australian supermarkets that emerged at a time when they were experiencing media and public criticism for their allegedly poor treatment of farmers and suppliers. During one integrated advertising activation developed in conjunction with a popular television cooking show, I observed that the Twitter conversation occurring during the episode broadcast replicated almost word-for-word the main advertising campaign's core brand messages, including the notion that farmers are happy and fulfilled both by their work and by their relationship with the major supermarket at the helm of the campaign.

Approaching the research question of "how is food work represented in media?" by starting with one text (*It Tastes Better*) and then moving outwards to consider other texts, including advertising and social media, revealed how almost identical imagery circulates with quite different politics and agendas. In one context, such as when associated with the small-scale producers of *It Tastes Better*, this imagery can be used to articulate an alternative food politics; in another, such as when used in the advertising campaign of a major supermarket, it can be used to "greenwash" the activities of major corporate players. Analysing the imagery in this way raises questions about how alternative food politics is typically mobilised in media and the ease by which its imagery and discourses can be appropriated for quite different goals. My study did not focus solely on digital food texts, but in locating textual representations in their movements across and between media forms, it reflects an approach to textual analysis necessary for food studies in a digital era. Crucially, it represents an approach where meaning is located not only "within" texts, but also in the relationships between them – something that is now essential as mediums and platforms becoming increasingly integrated and flows of content between them increasingly ubiquitous in the contemporary media landscape.

Key lessons for digital food studies

One of the key advantages of textual analysis is its broad applicability to a range of texts and contexts. In a rapidly changing media environment, textual analysis can be readily applied to new media texts as they emerge and evolve. But while traditional approaches to textual analysis have maintained enduring value in a digital era, they also require some updating. As the example I gave above suggests, capturing the complexity of digital food media demands moving beyond analysis of individual texts (such as a cookbook) to consider how repeated tropes and images move across and between media platforms, how they interact with other texts and genres (i.e., their intertexts), and how their textual conventions are shaped by industry conditions and imperatives.

Individual texts may invite particular interpretations, but they do not simply stand alone. To fully understand textual meanings, texts must be located within their specific political economies, and within their relevant cultural and political contexts. When analysing digital food texts, it is also essential to understand the affordances and "vernaculars" (Gibbs et al., 2015) of the online platforms under examination: that is, the specific potentials for action embedded in the hardware and software of different platforms, and the various narrative and communication patterns that shape content and information flows between them. The "platformisation" of media, where online applications are increasingly integrated with each other and with the wider web (Helmond, 2015), means that digital texts now move rapidly between platforms and contexts, and our interpretations need to be sensitive to these movements. For example, an online news story about vegan activism might be shared on Facebook, adapted for Instagram, ridiculed on a Reddit thread, and turned into an anti-vegan meme – the same initial content adding and changing meaning as it moves through these different contexts. Following representations of food as they shift between and across different platforms is often now essential for untangling the complex relationships between meaning, representation, and power in a digital era.

Using textual analysis: A how to guide

Using textual analysis in digital food studies requires careful attention to research project design. Because projects employing textual analysis as their primary method often do not need ethics approval or the recruitment of research participants prior to commencement, it can be tempting to skip the preparation and planning and get straight to work. However, a successful textual analysis requires researchers to think just as clearly about research design as those undertaking empirical methods.

When designing a study using textual analysis, the key questions to ask are: which texts? How many? How will you analyse them? Why are these choices the best ones for answering your research question? Careful justification of each step is required. Some research questions (e.g., "How does the 'clean eating' movement construct ideas about 'good' food?") may produce a very large number of texts for analysis, in which case you may consider using computational methods. However, it is important to keep in mind that when using textual analysis "more" is not necessarily "better." Larger samples may give you an overview of broad trends, but a well-chosen, smaller sample is much better for analysing deeper meanings. For example, if you are interested in analysing how the integrated advertising strategies of a television cooking show are received by online audiences, you might select a single television episode, and then follow this episode through its reception on news sites, social media, and online blogs. Such an approach will likely produce a deeper and richer analysis than attempting to analyse the show in its entirety, which would only allow you to barely "scratch the surface" of meaning.

Dos and Don'ts

The following "dos" and "don'ts" are helpful to consider when designing a digital food study using textual analysis:

Do make sure that your research questions are answerable using textual methods. Textual analysis can only tell us the meanings that a text *invites* (McKee, 2003); it cannot tell us how audiences (or media producers) actually interpret it. Textual analysis allows us to identify patterns of meaning and representation across texts, but understanding what these texts mean to those who produce and consume them requires further audience or industry study.

Do give careful consideration to your sampling and selection of texts. While it is theoretically possible to apply textual analysis to any number of texts, it is not necessarily advisable. Don't just choose texts that are convenient or that serve your argument: a clear justification of your selection, sourcing, and sampling strategies is necessary. Studies using textual analysis frequently require equally as rigorous an explanation of method as empirical approaches.

Don't allow your preconceptions of what you expect to find prevent you from reading texts on their own terms. Let the texts reveal to you their meaning. Many things that can feel instinctively true don't always hold up to the rigours of textual evidence.

Don't imagine that textual analysis is purely subjective or that any interpretation is possible. Interpretations always need to be supported by clear and compelling textual evidence, with texts appropriately located in their social, cultural, and industry contexts.

Now you try

Choose a food celebrity, and identify at least five media texts where the celebrity has appeared. Include a mix of different mediums and platforms, such as television, news, and social media. Taking into consideration the different production and consumption conventions shaping media representations in each of the mediums and platforms, how do the meanings associated with the celebrity change as they move between the different media? What meanings remain the same?

Questions for reflection

How do different research methods produce different kinds of knowledge about media texts? (e.g., what are the different way of thinking about and understanding media that are produced by discourse, framing, or content analyses?) How would you weigh up these differences when selecting an appropriate research method?

How would you respond to criticisms that textual analysis is not "objective" and its interpretations simply "opinion"?

What ethical considerations (if any) should be taken into account when conducting textual analysis on publicly available media texts, such as public websites? Would these considerations be different if you were analysing the website of a large food company compared to a small activist group?

Ethical considerations

Many text-based studies do not need formal ethics approval, though a number of institutions do now require ethical review for projects involving social media. But even in cases where ethical review is not required, it is important to consider the ethics of text-based methods as carefully as you would for any method. There are two key issues to consider:

Textual analysis is by its nature interpretive, and your interpretations may look quite different to those of media producers and consumers, who may find your interpretations confronting or challenging. While the intentions of producers or the views of consumers should not be taken as straightforward arbiters of a text's meaning, it is important to always locate your textual analyses in their proper contexts, attending to the purposes for which texts are made and the conditions under which they are produced and consumed.

If you wish to publish textual examples with your research, including images or quotes from social media, it is essential that you have a clear understanding of copyright and fair use. Some journals and book publishers will allow some material to be published under conditions of fair use; others will require permissions for *any* media materials used. Social media platforms often have specific (and varying) terms and conditions about how material from their sites can be used. Permissions can be expensive, and sometimes impractical, so it is important to plan for potential issues well in advance when designing studies involving textual research.

Further readings

Deacon, D., Murdock, G., Pickering, M., & Golding, P. (2007). *Researching Communications: A Practical Guide to Methods in Media and Cultural Analysis*. London: Bloomsbury.

Highfield, T. & Leaver, T. (2016). Instagrammatics and digital methods: Studying visual social media, from selfies and GIFs to memes and emoji. *Communication Research and Pratice*, 2(1): 47–62. doi: 10.1080/22041451.2016.1155332

McKee, A. (2003). *Textual Analysis: A Beginner's Guide*. London: Sage.

References

Andrews, M. (2003). Nigella bites the naked chef: The sexual and the sensual in television cookery programs. In J. Floyd & L. Forster (eds.), *The Recipe Reader: Narratives – Contexts – Traditions* (pp. 187–204). Aldershot: Ashgate.

Ashley, B., Hollows, J., Jones, S., & Taylor, B. (2004). *Food and Cultural Studies*. London: Routledge.

Barthes, R. (1977). *Image-Music-Text*. New York: Hill and Wang.

Bell, D. & Hollows, J. (2011). From *River Cottage* to *Chicken Run*: Hugh Fearnley-Whittingstall and the class politics of ethical consumption. *Celebrity Studies*, 2(2): 178–191. doi: 10.1080/19392397.2011.574861

Bell, D., Hollows, J., & Jones, S. (2017). Campaigning culinary documentaries and the responsibilization of food crises. *Geoforum*, 84: 179–187. doi: 10.1016/j.geoforum.2015.03.014

Buddle, E.A. & Bray, H.J. (2019). How farm animal welfare issues are framed in the Australian media. *Journal of Agricultural and Environmental Ethics*, 32(3): 357–376. doi: 10.1007/s10806-019-09778-z

Chan, A. (2003). 'La grande bouffe': Cooking shows as pornography. *Gastronomica: The Journal of Critical Food Studies*, 3(4): 47–53. doi: 10.1525/gfc.2003.3.4.46

Clancy, K.A. & Clancy, B. (2016). Growing monstrous organisms: The construction of anti-GMO visual rhetoric through digital media. *Critical Studies in Media Communication*, 33(3): 279–292. doi: 10.1080/15295036.2016.1193670

Contois, E. (2017). Healthy food blogs: Creating new nutrition knowledge at the crossroads of science, foodie lifestyle, and gender identities. *Yearbook of Women's History*, 36: 129–145.

Craig, G. (2019). Sustainable everyday life and celebrity environmental advocacy in *Hugh's War on Waste*. *Environmental Communication*, 3(6): 775–789. doi: 10.1080/17524032.2018.1459770

de Solier, I. (2005). TV dinners: Culinary television, education and distinction. *Continuum: Journal of Media & Cultural Studies*, 19(4): 465–481. doi: 10.1080/10304310500322727

Dejmanee, T. (2016). 'Food porn' as postfeminist play: Digital femininity and the female body on food blogs. *Television & New Media*, 17(5): 429–448. doi: 10.1177/1527476415615944

Entman, R.M. (1993). Framing: Towards clarification of a fractured paradigm. *Journal of Communication*, 43(4): 51–58. doi: 10.1111/j.1460-2466.1993.tb01304.x

Fairclough, N. (1995). *Media Discourse*. London: Edward Arnold.

Friedlander, J. & Riedy, C. (2018). Celebrities, credibility, and complementary frames: Raising the agenda of sustainable and other 'inconvenient' food issues in social media campaigning. *Communication Research and Practice*, 4(3): 229–245. doi: 10.1080/22041451.2018.1448210

Frow, J. (2006). *Genre*. London: Routledge.

Gallegos, D. (2005). Cookbooks as manuals of taste. In D. Bell & J. Hollows (eds.), *Ordinary Lifestyles: Popular Media, Consumption and Taste* (pp. 99–110). Maidenhead: Open University Press.

Gibbs, M., Meese, J., Arnold, M., Nansen, B. & Carter, M. (2015). #Funeral and Instagram: Death, social media, and platform vernacular. *Information, Communication & Society*, 18(3): 255–268. doi: 10.1080/1369118X.2014.987152

Hansen, A. & Machin, D. (2013). *Media & Communication Research Methods*. London: Palgrave Macmillan.

Harrington, S., Collis, C., & Dedehayir, O. (2019). It's not (just) about the f-ckin' animals: How veganism is changing, and why that matters. In M. Phillipov & K. Kirkwood (eds.), *Alternative Food Politics From the Margins to the Mainstream* (pp. 135–150). London: Routledge.

Helmond, A. (2015). The platformization of the web: Making web data platform ready. *Social Media & Society*, 1(2): 1–11. doi: 10.1177/2056305115603080

Henderson, J., Coveney, J., Ward, P., & Taylor, A. (2009). Governing childhood obesity: Framing regulation of fast food advertising in Australian print media. *Social Science & Medicine*, 69(9): 1402–1408. doi: 10.1016/j.socscimed.2009.08.025

Highfield, T. & Leaver, T. (2016). Instagrammatics and digital methods: Studying visual social media, from selfies and GIFs to memes and emoji. *Communication Research and Pratice*, 2(1): 47–62. doi: 10.1080/22041451.2016.1155332

Hollows, J. (2003a). Feeling like a domestic goddess: Postfeminism and cooking. *European Journal of Cultural Studies*, 6(2): 179–202. doi: 10.1177/13675494030060020003

Hollows, J. (2003b). Oliver's twist: Leisure, labour and domestic masculinity in *The Naked Chef*. *International Journal of Cultural Studies*, 6(2): 229–248. doi: 10.1177/13678779030062005

Hollows, J. & Jones, S. (2010). 'At least he's doing something': Moral entrepreneurship and individual responsibility in *Jamie's Ministry of Food*. *European Journal of Cultural Studies*, 13(3): 307–322. doi: 10.1177/1367549410363197

Hurwitz, L.B., Montague, H., & Wartella, E. (2017). Food marketing to children online: A content analysis of food company websites. *Health Communication*, 32(3): 366–371. doi: 10.1080/10410236.2016.1138386

Jackson, P., Russell, P., & Ward, N. (2007). The appropriation of 'alternative' discourses by 'mainstream' food retailers. In D. Maye, L. Holloway, & M. Kneafsey (eds.), *Alternative Food Geographies: Representation and Practice* (pp. 309–330). Sydney: Elsevier.

Jones, S. & Taylor, B. (2001). Food writing and food cultures: The case of Elizabeth David and Jane Grigson. *European Journal of Cultural Studies*, 4(2): 171–188. doi: 10.1177/136754940100400204

Ketchum, C. (2005). The essence of cooking shows: How the food network constructs consumer fantasies. *Journal of Communication Inquiry*, 29(3): 217–234. doi: 10.1177/0196859905275972

Kress, G.R. & van Leeuwen, T. (1996). *Reading Images: The Grammar of Visual Design*. London: Routledge.

Kristeva, J. (1980/1970). *Desire in Language: A Semiotic Approach to Literature and Art*. Trans. T. Gora, A. Jardine, & L.S. Roudiez. New York: Columbia University Press.

Lewis, S.C., Zamith, R., & Hermida, A. (2013). Content analysis in an era of big data: A hybrid approach to computational and manual methods. *Journal of Broadcasting & Electronic Media*, 57(1): 34–52. doi: 10.1080/08838151.2012.761702

Lewis, T. (2008). *Smart Living: Lifestyle Media and Popular Expertise*. New York: Peter Lang.

Lewis, T. (2011). 'You've put yourselves on a plate': The labours of selfhood on *MasterChef Australia*. In H. Wood & B. Skeggs (eds.), *Reality Television and Class* (pp. 104–116). London: Palgrave Macmillan/BFI.

Lewis, T. (2019). *Digital Food: From Paddock to Platform*. London: Bloomsbury.

Linné, T. (2016). Cows on Facebook and Instagram: Interspecies intimacy in the social media spaces of the Swedish dairy industry. *Television & New Media*, 17(8): 719–733. doi: 10.1177/1527476416653811

Lupton, D. (2017). 'Download to delicious': Promissory themes and sociotechnical imaginaries in coverage of 3D printed foods in online news sources. *Futures*, 93: 44–53. doi: 10.1016/j.futures.2017.08.001

Lupton, D. (2019). Vitalities and visceralities: Alternative body/food politics in digital media. In M. Phillipov & K. Kirkwood (eds.), *Alternative Food Politics From the Margins to the Mainstream* (pp. 151–168). London: Routledge.

Lupton, D. (2020). Carnivalesque food videos: Excess, gender and affect on YouTube. In D. Lupton & Z. Feldman (eds.), *Digital Food Cultures*. London: Routledge.

Mayes, C. (2014). Governing through choice: Food labels and the confluence of food industry and public health discourse to create 'healthy consumers'. *Social Theory & Health*, 12(4): 376–395. doi: 10.1057/sth.2014.12

McDonnell, E. (2016). Food porn: The conspicuous consumption of food in the age of digital reproduction. In P. Bradley (ed.), *Food, Media & Contemporary Culture: The Edible Image* (pp. 239–265). Houndmills: Palgrave Macmillan.

McKee, A. (2003). *Textual Analysis: A Beginner's Guide*. London: Sage.

Moseley, R. (2001). 'Read lads do cook...but some things are still hard to talk about': The gendering of 8–9. *European Journal of Cultural Studies*, 4(1): 32–39.

Pennell, M. (2016). More than food porn: Twitter, transparency, and food systems. *Gastronomica: The Journal of Critical Food Studies*, 16(4): 33–43. doi: 10.1525/gfc.2016.16.4.33

Phillipov, M. (2017). *Media and Food Industries: The New Politics of Food*. Cham: Palgrave Macmillan.

Philo, G. (2007). Can discourse analysis successfully explain the content of media and journalistic practice? *Journalism Studies*, 8(2): 175–196. doi: 10.1080/14616700601148804

Rodney, A., Cappeliez, S., Oleschuk, M., & Johnston, J. (2017). The online domestic goddess: An analysis of food blog femininities. *Food, Culture & Society*, 20(4): 685–707. doi: 10.1080/15528014.2017.1357954

Silver, J.J. & Hawkins, R. (2017). 'I'm not trying to save fish, I'm trying to save dinner': Media, celebrity and sustainable seafood as a solution to environmental limits. *Geoforum*, 84: 218–227. doi: 10.1016/j.geoforum.2014.09.005

Strange, N. (1998). Perform, educate, entertain: Ingredients of the cookery programme genre. In C. Geraghty & D. Lusted (eds.), *The Television Studies Book* (pp. 301–312). London: Arnold.

Thompson, K. & Haigh, L. (2017). Representations of food waste in reality food television: An exploratory analysis of *Ramsay's Kitchen Nightmares*. *Sustainability*, 9(7): 1–10. doi: 10.3390/su9071139

Vassallo, A.J., Kelly, B., Zhang, L. et al. (2018). Junk food marketing on Instagram: Content analysis. *JMIR Public Health and Surveillance*, 4(2). doi: 10.2196/publichealth.9594

Vidal, L., Ares, G., Machín, L., & Jaeger, S.R. (2015). Using Twitter data for food-related consumer research: A case study on 'what people say when tweeting about different eating situations'. *Food Quality and Preference*, 45: 58–69. doi: 10.1016/j.foodqual.2015.05.006

Wagner, K. (2015). Reading packages: Social semiotics on the shelf. *Visual Communication*, 14(2): 193–220. doi: 10.1177/1470357214564281

Weerakkody, N. (2009). *Research Methods for Media and Communication*. South Melbourne: Oxford University Press.

Yoo, J.H. & Kim, J. (2012). Obesity in the new media: A content analysis of obesity videos on YouTube. *Health Communication*, 27(1): 86–97. doi: 10.1080/10410236.2011.569003

Zukin, S., Lindeman, S., & Hurson, L. (2017). The omnivore's neighbourhood? Online restaurant reviews, race, and gentrification. *Journal of Consumer Culture*, 17(3): 459–479. doi: 10.1177/1469540515611203

3

READING CELEBRITY FOOD WEBSITES

Poststructural approaches

Katrine Meldgaard Kjær

An empirical interest in the digital seems inevitable for food studies researchers interested in the way we talk about and make sense of food in the contemporary. Social media, online activity, and apps shape the way we communicate and express ourselves in our everyday lives, not least in relation to food: recipes are now largely located online and on blogs, pictures of food are taken specifically to be shared on social media, and food recommendations and reviews are exchanged between peers via geo-specific apps. This chapter will provide tools for considering the specific type of digital platform of the website, and particularly websites connected to celebrity brands that deal with food and diets. The chapter will provide ways to consider how these websites are embedded in wider cultural norms and ideas, particularly through feminist perspectives on poststructural analysis, which focus on how bodily hierarchies shape and are shaped by the ways in which food and diet are represented and discursively constructed.

As celebrities' reach in a globalized world is both created and sustained via a presence on a multitude of digital platforms, contemporary celebrity is also becoming an increasingly digitalized phenomenon. While some celebrities are born on the digital platforms – such as YouTube stars and social media-based influencers – others have travelled from more analogue stages into the digital via a presence on social media and the Internet more generally. Regardless of their relationship to the digital, many celebrities have in recent years chosen to link their brand to food-related issues. With this link, they may capitalize on the current intense cultural attention paid towards food as forms of entertainment, escapism and as a key to health and wellbeing. Given the celebrity's position in contemporary society as an alternative type of authoritative voice on highly charged topics of all sorts (see Farrell, 2012), it is unsurprising that celebrities from a multitude of arenas are establishing themselves within the foodscape as well. Looking at celebrity brand websites is important to digital food studies, then, because it is one of the arenas where alternative voices of cultural authority on eating is being created, circulated, and profited from in the contemporary digitalized society.

This chapter will provide overviews of poststructural discourse analysis-based methodologies. A poststructural approach to digital food studies implies a focus on understanding not only on what something is, but also rather, how something comes to be, and what processes and relationships inform this becoming (Hvidtfeldt & Kroløkke, 2016: 16–17). With this, the chapter will focus on poststructural discourse analysis as a particular way to analyse food and diet on celebrity websites, with special attention towards locating the underling norms and logics that structure how food and diet are represented here.

Poststructuralism and discourse analysis

There are many examples of the mutually constructing relationship between celebrity and food, from the rise of celebrity chefs as moral entrepreneurs (see Hollows & Jones, 2010) to the mobilising of female domesticity in postfeminist image-creation (see Greene, 2005) to celebrities expanding their own brand by "doing food" (see Howard, 2015). As a case study, this chapter takes its point of departure in Gwyneth Paltrow's lifestyle brand goop.com, and the ways in which food is represented here (see Meldgaard Kjær, 2019). Goop.com heavily features food- and diet-related content, which have also become a cornerstone of Paltrow's lifestyle-image. The key role of food in goop and its connection to Paltrow's extended media production, in particular her cookbooks *It's All Good* (2013), *It's All Easy* (2016), and *The Clean Plate* (2019), has allowed Paltrow to transition from an acting career to a domestically focused lifestyle brand. At the same time, Paltrow's goop brand and its connection to luxurious practices and products has allowed her to capitalize on this transition, as she emerges via goop as "an embodiment of idealized whiteness (...) because she is discursively placed in proximity to objects and values which are culturally judged to be good" (Graefer, 2014: 110). Poststructural approaches to such a case are fundamentally interested in pulling apart and examining the ways in which meaning is attached to food in this context, and how that relates to contemporary power relationships or ideas about desirability and hierarchy.

The poststructural approaches presented in this chapter are a further development of structuralist theory, especially the structuralist linguistic theory of Ferdinand de Saussure. Saussure developed his theory as a critique of earlier ideas that language gains it's meaning by referring to reality outside of language itself. Instead, Saussure argues that the relationship between sign and signifier is random in principle but in practice governed by the conventions or structures of specific languages; this makes language intrinsically connected to the social. While poststructuralism draws on Saussure's basic idea that language is social, it also challenges structuralism's idea of stabile and schematic structures. Poststructuralist thinkers such as Jacques Derrida, for example argue that there will always be excess meanings that disrupt schematic, binary structures and displace its fixed meanings (Lykke, 2010: 100). The development from structuralism to poststructuralism thus centrally involves a focus on (open-ended) processes than on structures. Koch (2007: viii) argues that post-structuralism, at its most basic level, abandons the search for absolutes. This is also reflected in the ways in which method is conceptualized here, as poststructural approaches are generally characterized by not clearly distinguishing between distinct categorizes of method, theory and methodology (Esmark et al., 2005); rather, they rely on analytical strategies, which also are reflective and emphasize understanding how different approaches themselves create the object of inquiry, with different epistemic possibilities and problems (ibid: 10). This means that methodological guidance in relation to poststructuralist studies involves the development of questions the researcher can ask their object of inquiry to shed light on the internal logics of the object of enquiry and what they sustain. When applied to food studies, this orientation allows for a lens through which we can understand food not only as nutrition or a biological need, but indeed also examine how relationships to food and the discursive construction of food creates value, meaning, and power (see Lupton, 1996). A poststructural approach to studying food, then, is not primarily concerned with figuring out "facts" of and about food, but rather, how these "facts" are formed, understood, interpreted, negotiated, and resisted, and how they are a product of and produced within specific material, political, historical, and social contexts. A specifically feminist approach to this involves thinking about the specificities of cultural ideas about food, and how these play into the ways we understand gender, bodily hierarchies, and power position in society.

Language plays a key role in poststructural analysis, which "often implicitly or explicitly insists on language and reality, representation and reality and discursivity and materiality are impossible to separate" (Stormhøj, 2006: 151). For this reason, poststructural thought has been closely linked with discourse analysis as method. Defined broadly, discourse can be defined as "a particular way of talking about and understanding the world (or an aspect of the world)" (Jørgensen & Phillips, 2008: 1), and discourse analysis as the practice of analysing these ways of talking and understanding. Discourse analysis, however, is also an umbrella term for multiple methodologies and ways to approach this analysis. These approaches are often rooted in approaches based on a specific theorist, for example, Ernesto Laclau and Chantal Mouffe (1985), Michel Foucault (1972), or Norman Fairclough (1989), each of which has specific ways of theorising discourse and methods for analysing it. Drawing on Burr (1995) and Gregen (1985), Jørgensen and Phillips (2008) identify four key premises that the diverse field of approaches to discourse analysis share, most notably on account of being broadly rooted in social constructionism: 1) that our knowledge and representations of the world are not objective, but a product of discourse; 2) that the ways in which we understand the world are historically and culturally contingent; 3) that knowledge is created through social interaction in which we construct common truths; 4) that different social understandings of the world leads to different social actions, and that the social construction of knowledge and truth therefor has social consequences (5–6). It is thus a basic idea in discourse analysis that our way of understanding the world is filtered and shaped through discourse, which is why it is an important object of study, as well as a site for critical intervention.

This chapter draws especially on Foucauldian discourse analysis traditions, which emphasise power relationships and the history of how discourse can be used as an instrument in compliance to certain norms about, in particular, "right" or "wrong" types of bodies. With these focus areas, Foucauldian approaches have enjoyed particular prominence within feminist research. A key figure in feminist poststructural analysis is Judith Butler (1990, 1993) who draws especially on Foucault and speech act theory to develop a theorisation of gender as constituted discursively, and being performative, as gender becomes naturalised through "continuous repetition and citations, norms about "right/wrong" or "natural/unnatural" ways of doing or performing gender" (Lykke, 2010: 91). Here, discourse is tied to categories of assumed truths about appropriate gender behaviour, which shape our ideas about gender and how we speak about it, and "perform" it. Keeping with the idea that discourse has social power, it is central to this argument that the ways we speak about gender takes on social significance because restricts and limits which types of gendered behaviour can be deemed acceptable or "natural." Butler's work has been highly influential in feminist scholarship, as it draws attention to how (gendered) ideas about the body can be tied to norms for the "right" or "wrongness" of an individual subject, and how speech acts and discourse is a vehicle for this.

Methodologically, scholars of discourse studies have argued that discourse analysis approaches share a basic assumption that discourse analysis deals with the three "components" of language, practice, and context, in the interplay of which discourse emerges (Angermuller, Maingueneau, Wodak, 2014: 6). While one of these three components in often the empirical focus of the analysis, and the other two might have a more theoretical character, but it is characteristic of discourse analysis that all three components much be present in the analysis (ibid.). These three components are thus central to how discourse analysis can be a way to operationalise poststructuralism's concern with underlying logics; discourse can be studied by examining how language is related to contexts and practices of a given specific case, which can in turn enable us to see how discourse produces knowledge and power in these cases. This triangle of language, practice, and context is thus also what enables the critical potential of discourse analysis.

Reading websites as texts

As a digital platform, a website can be studied in a multitude of ways, each with its own set of specific challenges and implications. This chapter is concerned with discourse analysis research into websites, which focuses on close and contextual readings, requires only access to the website in question, may be done manually and without digital tools or software. This differs from the use of computational and big data-based methods for studying websites which have gained popularity over last few decades within the fields of digital humanities and digital methods (see e.g., Rogers, 2019). Case- and circumstance specific work may be difficult to conduct, however, if relying more on quantitative or purely digital methods, which often remove data from its original context and into larger datasets; for discourse analysis focused on the interrelation between language, practice and context, then, it makes critical sense to rely on close, qualitative readings of the website.

Goop.com is a website, but it is also a portal, an entry-point to an array of sub-topics within the website universe of more general lifestyle advice (i.e., https://goop.com/food/ or https://goop.com/wellness/). Goop.com is a collection of multiple sub-fields of content, from fashion sections to advertisements to travel guides to recipes and diet advice, all of which are curated to fit the goop brand, and by extension, Gwyneth Paltrow persona. The website is essentially a set of ordered and catalogued collections that have cultural significance and are accessible to the public. Therefore, we may also view this type of website as a type of archive (Gale & Featherstone, 2011: 19); an archive of lifestyle texts as it is imagined in the contemporary moment of late capitalism. For poststructural theorists such as Michel Foucault (1972) and Jacques Derrida (1996), archives are home to a multiplicity of discourses, which are interesting to study because they are textual reconfigurations of the power dynamics of its contemporary time (Featherstone, 2006: 596; see also Bowker, 2010).

Indeed, studying discourses on a website involves approaching this website as a text. Web historian Niels Brügger (2010) argues that as an analytical object, the website should first and foremost be seen as a mediated textual artifact, that is, textual units within the medium of the internet (p. 33). It is a basic assumption within cultural studies is that every cultural practice or product can be analysed as text which may be "read" and interpreted – this is also true food related practices on websites. As argued in Leer (2014), semiotics scholars such as Roland Barthes (1979) have paved the way for cultural food analysis of media products by showing that textual analysis can be applied to many different types of texts and be used to understand and analyse nuances and significances in these. Within cultural studies and in the tradition of Roland Barthes, textual analysis has generally been used as a tool to examine underlying cultural assumptions of a text and the cultural context in which it is situated (Fürsich, 2009: 240). These themes are studied via close readings of the way food is written about on the website, both as a whole and in specific posts, articles, or entries, and by relating this to the cultural or societal context in which it is situated. Keeping in mind a broad definition of "text," the analysis is not only confined to words, but also visuals and sounds such as pictures, illustrations, videos, music, etc.

Discourse analysis of a website involves close readings, and is often based on specific case studies, such as a single website and the celebrity brand this is connected to. Although case studies are sometimes accused of being too narrowly focused on a specific phenomenon, Bent Flyvbjerg (2006) argues that case studies can provide generalisable findings even as, or perhaps precisely because, they focus on concrete, context-dependent knowledge (p. 223). A dataset collected for a close reading of a specific case study may thus consist of material from a single website. Since the focus is on an in-depth examination of a single case, the data collection from this case should aim at collecting the many different aspects of the website "text," as these extend to not only both written articles and comments, but also the text of images and videos, webpage design, menus, and the organisation of "texts," broadly speaking, into categories or collections on the website. These are all a part of the context in which the single textual elements that can be analysed individually are embedded, and all say something about the ways in which food is represented on a given website, including the ways in which it is connected to a particular celebrity's image. Because the goop.com project was based in a feminist discourse analysis tradition, my initial reading strategy consisted of looking for an identifying assumed or naturalised beliefs or statements about the body in both textual and visual material (how was dieting written about in relation to the body, how were bodies visually represented in relation to food); identifying how imagery and text together conveyed norms for the body and its relationship to food (which types of behaviour or ideas around food and the body were assumed natural, necessary, or normal in text and visuals); and how the layout of the website furthered or supported the messages in the textual content (how did the design of the website support or accentuate specific texts or visuals about the body and dieting). This initial reading strategy is an example of relatively traditional reading strategies within critical poststructural discourse analysis. In recent years, however, traditional poststructural discourse analysis methods have been critiqued within feminist scholarship for not being adequately concerned with materiality (Barad, 2003; Alaimo & Hekman, 2008). Despite having the body as a central object of concern, scholars within what is often called new materialism or feminist materialism argued that feminist poststructural approach to discourse analysis was often insufficient in its ability to engage with the materiality of body and the many other materiality that surround and constitute it. This critique is also central when thinking about how the materiality and technical aspects of the website can be accounted for in poststructural reading strategies.

From discourse to materiality

In a case study of the national Danish broadcasting service, DR (Denmark's Radio), historian Helle Standgaard Jensen (2017) argues that the interface, construction and content of an electronic archive has a fundamental impact on the sources that are found in it and the stories these sources can tell (p. 69). Therefore, she argues, these elements must be considered active participants and co-constructers of what is "read" and analysed. This is very much in line with the increased attention towards materiality in feminist poststructural theory in recent years, which calls for more consideration of how techno-material factors impact the texts that are "read" in poststructural analysis, as well as with earlier work within fields such as Actor-Network Theory (ANT) and Science & Technology Studies (STS), in particular from scholars such as Bruno Latour and Donna Haraway. Both scholars highlight non-human actors, or materiality, as central in shaping cultural phenomenon, and argue that we cannot analyse such things without paying attention to them. Moreover, an analytical attention inspired by ANT and particularly STS scholarship focuses on the practices that shape the management and the governance of the

Internet, and determine the ways in which it operates, works, resists, and functions (Musiani & Schafer, 2019: 75).

In relation to studying websites, this scholarship contributes with a perspective on how a given website's content does not exist in a vacuum; rather, this content is a product of technical affordances, characteristics and properties and agencies, and those affordances interact with and shape human action in maintaining the archives. One specific way in which this was relevant to the goop.com project was that the materiality of website shifted multiple times during my research; not uncommonly for websites, the interface changed over time, and articles and functions which were present at one point could be gone a few months later. The aesthetics, visuals, and overall look of the archive also changed or were updated over time – a challenge to a research project working primarily with discourse analytical methods, which are dependent on somewhat stabilised content to "read." Discourse analysis requires that content be stabilised; it provides an analysis based on a snapshot in time of the website. But the website is rarely a stabile phenomenon, as it is often constantly updated, edited, and changed. In this way, websites differ from more traditional sources that textual discourse analysis methods are used to unpack, such as books or films. The data that one deals with when working with websites are lively (Lupton, 2016) and they are often in movements beyond the researcher's control where content shifts, disappears, and reappears. When working with websites, documentation thus becomes important, as does a reflexive consideration of the effects of this documentation and stabilisation; reading a "snapshot" in time requires that it is transparent to readers where this information and text comes from, as they may not be able to find it themselves online when the research is published. In the documentation process, it can be helpful to draw on other qualitative traditions such as ethnography, that have developed methods for writing fieldnotes and observations, also in relation to the digital (see e.g., Kozinets, 2015). Taking notes of what you observe on the website every time you visit it, as well as keeping a daily log of when and what is changed can be key in being able to decipher and thereby also analyse the changes and what they may mean for the way in which the website acts as a cultural archive. Moreover, a simple solution of extensive screenshots of the website can be helpful. Screen recording tools, many of which are open source and free, can also allow for the capture the website "live," including the ways in which you scroll through and use the site.

The theoretical "turn" towards materiality in feminist poststructural literature also involves a critique of overreliance on only studying human speech. A key actor in the feminist materialist critique of poststructural discourse analysis has been Karen Barad (2003), who challenges especially Butler's work for its focus on language. Barad does not disavow discourse or language entirely, but nonetheless argues that poststructural analyses need to be able to take other factors than language into account to understand how phenomena are created. This challenges the basic idea of discourse analysis, as she argues that the non-human and materiality should be more in focus in these analyses than they have traditionally been. Likewise, other scholars working with materiality in poststructural analysis highlight the need to incorporate into these analyses how "various aspects of materiality contribute to the development and transformation of discourse" (Alaimo & Hekman, 2008: 2). As Lupton (2018) argues, feminist materialism develops poststructural analytical modes by offering a theoretical perspective that acknowledges the role played by nonhumans such as digital media and devices. For digital food studies, this approach implies a focus on the interaction between humans and non-human objects of food, as well as how the technologies they are embedded within, are constantly in a co-constructing relationship, where each is configured and reconfigured (ibid.).

The process of accounting for materiality is an ongoing discussion in feminist poststructuralist theory. Much work here is of a theoretical-methodological nature that reflects poststructuralism's traditional rejection of absolutes and lack of clear divisions between theory and method. However, in my work with goop.com, I found simple mapping techniques to be a helpful and tangible starting point to develop a reading strategy that involved materiality, one example being Adel Clarke's (2005) situational analysis mapping. Situational analysis mapping is a form of flexible coding or sorting of the elements of the a given text that highlights the relationship between human and non-human actors, where initial "messy maps" focus on laying out "as best one can all the most important human and non-human elements in the situation of concern of the research broadly conceived" (p. 7), and the following "sorted map," is created by mapping the relationships between these actors. Mappings such as this which focus on materiality and the non-human, then, can make clear and explicit the relationships between different elements and actors in any given narrative – which can be difficult to have an eye for when coming from a tradition that looks primarily at language. From this, questions for reading strategies can then be developed. This reading strategy can revolve around asking questions to uncover how materiality informs how power is distributed and enacted in the text (Lupton, 2020), such as, how does materiality and the non-human disrupt, affect, or change the text? What might a purely language-focused approach miss in relation to this? It can also help to enable a consideration of how non-human actors and their materiality inform how ideas and norms about food and diet are configured and enacted on websites. This type of reading strategy thus widens the idea of discourse analysis to also include perspective of actors that challenge or shape discourse itself. In this way, it is a development of the focus on (open-ended) processes that is characteristic of poststructural approaches more generally.

Dos and Don'ts

In this chapter, I have outlined analytical approaches that focus on unpacking how cultural norms and assumptions are represented on celebrity brand websites about food. Poststructural approaches to discourse analysis do not typically distinguish clearly between theory and method, and are therefore often characterised by a more theoretical nature than other methodological descriptions. However, I have outlined how the study of food-related websites may be approached as texts and products of technical infrastructure, and have highlighted feminist materialist critique of discourse analysis' limited focus on language. I have proposed central questions that may guide a more materiality-focused reading and have highlighted how discourse analysis might draw on the methodological insights from other qualitative traditions for tools to structure reading strategies. The approaches I have introduced in the chapter are especially suited for enabling knowledge about small-scale, context specific case studies. Therefore,

Do situate the research in its specific contexts at the time of the research being carried out. What is the cultural function of the website or digital archive, and which dynamics have contributed to its creation and status?

Don't forget to document, whether it be in the form of screenshots, video, and/or fieldnotes. Websites are living, changeable entities.

Do reflect on what the form of context stabilisation you chose (screenshots, videos etc.) has on the ways in which you see and interact with the content when studying it: what is gained and lost in the stabilisation or preservation of the analysed content? What effects does stabilising otherwise lively content have?

Don't limit yourself to a predetermined idea of what kinds of relationships you will see in your analysis: working with feminist discourse analysis means working empirically and interpretatively with identifying subtle, surprising, and hidden "truths" as these appear in the text.

Do consider how the technical infrastructure of the text or archive shape the possibilities for its use, and for the research that may be conducted within it, specifically when using discourse analysis methods.

Now you try

Identify a celebrity lifestyle website that you would like to examine closer. This could be a website that you have used yourself or that you have heard of through other media sources. Visit the website and take an initial look through it, noting whatever what strikes you in your first encounter with this website: is there anything unusual or unique about it, or does it remind you of other types of websites in this category? Pick a few other websites that you can compare it to, and spend some time reading and researching the celebrity brand that is associated to the website you have picked. Notice how the celebrity is mobilised or related to in the different websites, and consider what is particular about the ways in which the celebrity is mobilised in your website of choice.

Revisit the website and investigate the origin of and power dynamics behind the website, including how was it started, by whom and why. Mapping out where the website comes from will help you understand the context within which it is embedded, and how representations of food here reflect that. Start collecting your dataset by documenting the diverse kinds of food-related texts and visuals on the website. Now look over the material you have collected and consider what is at stake here. When doing a feminist analysis, questions of norms and normative beliefs about gender and bodies, and how these are related to inclusions and exclusions, are particularly central when considering the stakes of a text. Accordingly, consider how the material you have collected from your case can help shed light on this in relation to issues of food and eating, and articulate one or more questions that can continue to guide your analysis of the material you have collected. Consider also how the technical infrastructure and makeup of the website are impacting the ways you may answer these questions, including how your analysis is structured by the affordances of the website, and how the website's make-up shapes the analysis you can make. As you work through the collected material, you will need to engage in an ongoing reflexive process of asking yourself how your reading strategy is shaping your analytical results, and what you are missing; analytical strategies are not static but should rather be seen as living beings that are developed in constant dialogue with the research project's theoretical framework.

Questions for reflection

How is digital media used in framing food as a part of contemporary celebrity brands? Which types of arenas do food advocacy anchor celebrities in, and what may be the pros and cons for celebrities to engage in this work?

In which ways may digital representations of diets tell us something about our contemporary relationship to food and body? In which ways are we seeing a push back or resistance to this?

Considerations about archives have traditionally been placed within disciplines related to history and the past. What may be gained from applying the concept to contemporary cases about digital food? What may be lost?

Ethical considerations

Discourse analysis is interpretative, and it is important to consider your positionality as researcher when doing this work. This positionality could be in relation to how your own background and research goals and questions may impact your readings, as well as a consideration of what is "left out" in your analyses. Analytical considerations should always include an explicitisation of the position from which you are reading and interpreting.

When relying on material from websites, it is important to consider issues of personal data and privacy. This involves a consideration of whether the material you are studying is personal or private, and how you can or should protect that privacy, as is accessing the public character of the material and the ethics of writing about should follow. Ethical considerations should be specific to the material, and may, for example, differ when looking at a highly public and professionalised celebrity website and a private citizen's personal blog.

Further readings

Brügger, N. (2010). *Web History*. New York: Peter Lang.
Jørgensen, M. & Phillips, L. (2008). *Discourse Analysis as a Theory and Method*. Thousand Oaks: Sage.
Lykke, N. (2010). *Feminist Studies. A Guide to Intersectional Theory, Methodology and Writing*. London: Routledge.

References

Alaimo, S. & Hekman, S. (eds.). (2008). *Material Feminisms*. Bloomington and Indiapolis: Indiana University Press.
Angermuller, J. Maingueneau, D., & Wodak, R. (2014). *The Discourse Studies Reader*. Amsterdam: John Benjamins Publishing Company.
Barad, K. (2003). Posthumanist performativity: Toward an understanding of how matter comes to matter. *Signs: Journal of Women in Culture and Society*, 28(3): 801–831.
Barthes, R. (1979). *Mythologies*. Boulder: Paladin.
Bowker, G.C. (2010). The archive. *Communication and Critical/Cultural Studies*, 7(2): 212–214.
Burr, C. (1995). *An Introduction to Social Constructionism*. London: Sage.
Butler, J. (1990). *Gender Trouble*. London: Routledge.
Butler, J. (1993). *Bodies That Matter. On the Discursive Limits of Sex*. London: Routledge.
Clarke, A.E. (2005). *Situational Analysis*. Thousand Oaks: Sage.
Derrida, J. (1996). *Archive Fever: A Freudian Impression*. Chicago: Chicago University Press.
Esmark, A., Laustsen, C.B., & Andersen, N.Å. (2005). Poststrukturalistiske analysestrategier. En introduction. In Esmark, A., Laustsen, C., & Andersen, N. (eds.), *Poststrukturalistiske analysestrategier* (pp. 7–41). Copenhagen: Samfundslitteratur.
Fairclough, N. (1989). *Language and Power*. London: Longman.
Farrell, N. (2012). Celebrity politics: Bono, product (RED) and the legitimising of philanthrocapitalism. *The British Journal of Politics and International Relations*, 14(3): 392–406.
Featherstone, M. (2006). Archive. *Theory, Culture & Society*, 23(2–3): 591–596.
Flyvbjerg, B. (2006). Five misunderstandings about case-study research. *Qualitative Inquiry*, 12(2): 219–245.
Foucault, M. (1972). *The Archaeology of Knowledge*. New York: Pantheon Books.
Fürsich, E. (2009). In defense of textual analysis. *Journalism Studies*, 10(2): 238–252.
Gale, M.B. & Featherstone, A. (2011). The imperative of the archive: Creative archive research. In B. Kershaw (ed.), *Research Methods in Theatre and Performance* (pp. 7–40). Edinburgh: Edinburgh University Pres.
Graefer, A. (2014). White stars and orange celebrities: The affective production of whiteness in humorous celebrity-gossip blogs. *Celebrity Studies*, 5: 107–122.
Greene, C.P. (2005). The domestic goddess postfeminist representation in the televisual kitchen: A media ecological analysis of Nigella bites. *Explorations in Media Ecology*, 4(3–4): 259–280.
Gregen, K. (1985). The social constructionist movement in modern social psychology, *American Psychologist*, 40(3): 266–275
Hollows, J. & Jones, S. (2010). 'At least he's doing something': Moral entrepreneurship and individual responsibility in Jamie's *Ministry of Food*. *European Journal of Cultural Studies*, 13: 307–322.
Howard, E. (2015). From fasting toward self-acceptance: Oprah Winfrey and weight loss in American culture. In J. Watson & E. Harris (eds.), *The Oprah Phenomenon* (pp. 101–124). Lexington: The University Press of Kentucky.

Hvidtfeldt, K. & Kroløkke, C. (2016). *Sundt, sundere, helt sygt. Kulturanalytiske tilgange.* Copenhagen: Samfundslitteratur.
Jensen, H.S. (2017). Digitale arkiver som medskabere i historieskrivning. In K. Drotner & S.M. Iversen (eds.), *Digitale metoder. At skabe, analysere og dele data* (pp. 69–87). Copenhagen: Samfundslitteratur.
Jørgensen, M. & Phillips, L. (2008). *Discourse Analysis as a Theory and Method.* Thousand Oaks: Sage.
Koch, A. (2007). *Poststructuralism and the Politics of Method.* Lanham: Lexington Books.
Kozinets, R. (2015). *Netnography Redefined.* Thousand Oaks: Sage
Laclau, E. & Mouffe, C. (1985) *Hegemony and Socialist strategy. Towards a Radical Democtratic Politics.* London: Verso.
Leer, J. (2014). *Ma(d)skulinitet. Maskulinitetskonskruktioner I madprogrammer efter The Naked Chef i lyset af "den maskuline krise".* PhD dissertation, University of Copenhagen, Copenhagen, Denmark.
Lupton, D. (1996). *Food, the Body and the Self.* Thousand Oaks: Sage.
Lupton, D. (2016). Personal data practices in the age of lively data. In J. Daniels, K. Gregory, & T. McMillan Cottom (eds.), *Digital Sociologies* (pp. 339–354). Chicago: Chicago University Press.
Lupton, D. (2018). Vitalities and visceralities: Alternative body/food politics in new digital media. In M. Phillipov & K. Kirkwood (eds.), *Alternative Food Politics: From the Margins to the Mainstream.* London: Routledge.
Lupton, D. (2020). Vital materialism and the thing-power of lively digital data. In D. Leahy, K. Fitzpatrick, & J. Wright (eds.), *Social Theory, Health and Education* (pp. 71–80). London: Routledge.
Lykke, N. (2010). *Feminist Studies. A Guide to Intersectional Theory, Methodology and Writing.* London: Routledge.
Meldgaard Kjær, K. (2019). Detoxing feels good: Dieting and affect in 22 Days Nutrition and goop detoxes. *Feminist Media Studies*, 19(5): 702–716.
Musiani, F. & Schafer, V. (2019). Science and technology studies approaches to web history. In N. Brügger (ed.), *The Sage Handbook of Web History* (pp. 71–85). Thousand Oaks: Sage.
Rogers, R. (2019). *Doing Digital Methods.* Thousand Oaks: Sage.
Stormhøj, C. (2006). *Poststrukturalismer. Videnskabsteori, analysestrategi, kritik.* Copenhagen: Samfundslitteratur.

4

ANALYSING DIGITAL FOOD SOUNDS FROM A TEXTUAL PERSPECTIVE – A CASE OF CHAMPAGNE (?)

Nicolai Jørgensgaard Graakjær

Introduction

Let us begin by listening to the sounds from this video distributed by *Epidemic Sound* (see Epidemic ASMR, 2017). Then take a moment to reflect on: What did you listen to?

At first glance, you might answer "sounds of champagne," possibly inspired by the video's title ("Champagne Bottle Open Sound"). However, upon closer inspection, you might question whether you can in fact hear from the sequence of sounds alone, that you are listening to specifically champagne and not, for example, another type of sparkling wine. This question is particularly relevant when sounds appear in digital media settings, where you are not able to determine the actual origin(ality) of the sounds you hear. This could then lead you to examine the sounds "themselves," that is, the structure and quality of the sounds and what they actually sound like.

This case illustrates the existence of an auditory dimension to digital media's representation of our interaction with foods. The case is of course not representative of all dimensions and variants of digital food sounds. However, the case represents a well-known and exemplary type of sound in an accessible format. The latter feature is particularly relevant for the present purposes, since the analysis of (digital food) sounds cannot rely on "stills" or "frozen sound events" but must be based on "repeated hearings of a single sound" (Chion, 1994: 32). Furthermore, the case illustrates how digital media – here exemplified by YouTube – allows for new practices of producing, distributing, and using (food) sounds. Specifically, the case video demonstrates, how a particular sequence of non-musical and non-verbal sounds has been "distilled" and focussed upon for users to comment on, redistribute, and enjoy (or examine) at their convenience.

The case will be used to demonstrate how to analyse digital food sounds from a textual perspective. In doing so, the chapter seeks to present text analytical concepts and strategies that can be generalised to the examination of other examples of digital food sounds. The sounds of foods are potentially highly significant, as they can produce food related information and influence our perception of foods even if we are not able to *see* the food in question in digital media settings like films and commercials. The chapter will exemplify, how food sounds do not only indicate the particular type of food from which the sounds would seem to originate. The sounds also inform us about the quality of the foods, the activities or processes of the handling the foods (food production, preparation, and/or consumption), and the type of social and physical setting of the handling of the foods. The chapter can thus hopefully help inspire a nuanced and systematised analysis of the significance and influence of sounds in digital food studies, where visual and verbal dimensions usually receives the most attention.

The chapter's text analytical approach is informed by semiotics as it aims to show how sound embodies part of the potential meanings that can be considered to be built into digital foods (inspired by Danesi, 2007: 3). Specifically, the chapter is stimulated by Chion's (1994) identification of three "modes of listening" (introduced in the context of film sound) as well as Peirce's tripartite classification of modes of relationships between a representamen and its object (or signifier and signified, following Chandler, 2007: 36). Whereas the three modes of listening can help organise and sharpen the auditory analytical attention, the modes of relationships can help conceptualise and describe ways in which sounds produce potentials of meanings (for more on the relationships between the perspectives of Peirce and Chion in the context of sound analysis, see Capeller, 2018).

The chapter begins by positioning the case as a particular variant of food sounds. Next, the chapter focusses on the case sounds and demonstrates how to examine them from a textual perspective. The chapter concludes by suggesting additional analytical perspectives.

Digital food sounds

Digital food sounds are here broadly defined as sounds that are somehow connected to the representation of foods in digital media. The case represents a particular variant of the general category of food sounds.

Firstly, sounds can be differentiated on the basis of a commonly accepted tripartite distinction (van Leeuwen, 1999) between music, speech, and a third residual category here referred to as object sounds. Although object sounds might occasionally be appreciated as – or even incorporated into (see below for an example) – music, the sounds do not qualify as music as they do not embody a discernible melodic and/or rhythmic structure. Also, even though the sounds could seem to "speak to" us, they are clearly distinguishable from speech as they are non-verbal and not produced by the human voice. From the perspective of types of sounds, the case hence illustrates object sounds.

Secondly, inspired by available classifications (see, e.g., Schifferstein, 2016), foods can be fleshed out by distinguishing between different food activities or processes.

In this context, I distinguish between the sounds of food production, preparation, and consumption. From this perspective, the case illustrates sounds from the preparation of foods.

Thirdly, inspired by the distinction between "sounds of objects" and "sounds for objects" proposed in the context of sound branding (Graakjær & Bonde, 2018) as well as the distinction between "food-intrinsic factors" and "food-extrinsic factors" in the context of food studies (Wang et al., 2019), I suggest that the general category of "food sounds" should be divided into three subcategories. I propose the term "sounds of food" to refer to sounds that originate from – and thus emerge as an intrinsic "property of" – the particular substance of food alongside other intrinsic factors such as "colour, aroma, texture, viscosity" (Wang et al., 2019: 1). "Sounds at food" is used to account for sounds that originate from objects in the "immediate vicinity" of producing, preparing, and/or consuming the food substance, for example, from its packaging or cutlery. I suggest the term "sounds with food" to address sounds that originate from "outside" the food substance and appear as an accompaniment to it. The distinction between "at" and "with" represents a distinction of what have previously been suggested to belong to the same category of extrinsic factors, that is "properties of product packaging or servingware, background music, ambient lighting, temperature and aroma" (Wang et al., 2019: 2). From the perspective of sound analysis, there is a close(r) relationship between the sounds of the food substance and its packaging or cutlery compared to the relationship between the substance and the – possible but not necessarily food related – accompanying music and speech. Consequently, object sounds of food constitute "the core" of food sounds. By adopting a text analytical approach focussed on an example of "sounds of food," the chapter supplements the mainly quantitatively oriented interest for "sounds for food" or extrinsic factors – occasionally termed "sonic seasoning" (Spence & Wang, 2015: 7) – mostly focussed on music (e.g., Spence & Wang, 2015a, 2015b; Spence et al., 2019).

In the following sections, the chapter thus focusses on an example of sounds of food. Specifically, the sounds from 0:15 to 0:30 in the video (mentioned at the beginning of the chapter) will be used to demonstrate how to analyse digital food sounds from a textual perspective. The demonstration is organised according to the three modes of listening and begins by addressing reduced listening which is arguably the most overlooked of the three modes. Subsequently, the chapter examines the case from the perspectives of casual listening and semantic listening. As will become clear, the three perspectives overlap and combine, but they are presented separately for educational reasons.

Reduced listening and iconic signification

From the perspective of reduced listening, the analysis addresses what the sounds *sound like*.

According to Chion, reduced listening: "takes the sound [...] itself the object to be observed instead of as a vehicle for something else" (1994: 29). However, in this context, the study of the sounds "themselves" will serve as the springboard for an examination of iconic signification – that is, a relationship between signifier and signified based on imitation or resemblance (Chandler, 2007: 36). Although we are habitually inclined to listen to sounds from the perspective of what they originate from (see more below), reduced listening has the advantage of "opening up our ears and sharpening our power of listening" (Chion, 1994: 31). Consequently, reduced listening facilitates a specification of the distinctive features of the sounds of particular foods.

The analysis of the sounds' quality and structure includes an analysis of "some distinctive, recognizable, and recurring physical form" (Danesi, 2007: 29). Accordingly, the temporal and dynamic form of sounds can be analysed from the perspective of *sound envelope*, which specifies a sound events' attack, decay, sustain, and release (Tagg, 2013: 277f). The case sounds (from 0:15 to 0:30) actually include several distinct sound events, including a pop sound followed by coexisting sounds which might be transliterated as "fizz" and "dluk, dluk, dluk." The pop sounds (at 0:15) embodies an abrupt and forceful attack with no extended decay, sustain nor release. By comparison, the fizzing sounds evolve through a prolonged attack (the sound initially emerge with an increase of intensity from 0:16 to 0:27) followed by a short-lasting decay (a decline in the intensity of sound at 0:28) which sustains only very shortly (regarding the continuation of the sound) before a fading release (the ending of the fizzing sound from 0:29 to 0:30).

Further aspects of what the sounds "sound like" can be approached from the perspective of sound modality, as presented by van Leeuwen (1999). Modality refers to what extent sounds embody "a true representation of the people, places and/or things represented" (ibid.: 180). Van Leeuwen offers eight articulatory parameters (1999: 172ff) for the analysis of degrees of modality. Already, the parameters *durational variety* and *dynamic range* have been included when examining the case sounds from the perspective of envelope. Additionally, the *pitch range* of the case sounds is moderate, and embody, for example, a minor increase in the pitch (or "melody height") of the fizzing sounds from 0:19 to 0:22. There is no significant *degree of friction* (e.g., the sounds are not distorted), whereas the fizzing sounds embody some *degree of fluctuation* (vibrato) through the variations in the sounds of a myriad of concurrent sonic microevents (bursting bubbles). From these articulatory parameters, the case sounds resemble – and hence have an iconic relationship to – the sound sequence produced by opening and pouring a bottle of carbonated liquid. The case sounds do not quite embody a "true" representation, though. This becomes clear, when the sounds are examined from the perspective of the remaining three parameters.

As regards *perspectival depth*, there is no differentiation of background and foreground. The sound sequence is "all alone" and it appears as "figure" without a "ground" (van Leeuwen, 1999: 181). As to the *degree of directionality*, the sounds appear in extreme "close up," as if we were listening to the sounds with our ears in very close proximity to object. Moreover, *absorption range* is low, as the sounds appear without reverberation – the general quality of the sounds are thereby, somewhat paradoxically, "dry" (van Leeuwen, 1999: 181). The sounds are intensified and decontextualised, and they appear hyperbolic, that is, somewhat "exaggerated." These features can be seen to reflect the context of the sounds as they are made available for a form of reduced and focussed listening with the sole purpose to relax and tingle the listeners thereby offering them a so-called autonomous sensory meridian response (or ASMR; see Lewis, 2019: 65). This purpose corresponds to the *sensory coding orientation* which listeners can adopt when they judge the modality of sounds and where "what matters is emotive effect, the degree to which the sound event has an *effect* of pleasure or its opposite" (van Leeuwen, 1999: 179). Furthermore, the hyperbolic nature of the sounds could be considered to mirror conventions of advertising, where brand objects are typically represented "looking (or, indeed, sounding) their best" and not necessarily the way they will precisely appear in all settings of real life. Van Leeuwen provides the related example of a lemonade commercial, where the "bubbles fizz and the ice cubes clink against the glass in crystal-clear and hyper-real fashion" (1999: 176).

In addition to "sound like" the specific sound sequence of the opening and pouring of a bottle of carbonated liquid, the case sounds resembles the structure and quality of other sounds and phenomena. The concept of anaphones – offered by Tagg (2013: 486ff) in the context of music analysis – can arguably inspire the examination of such resemblances. In this context, anaphones are used to address the case sounds' possible homologous relations – representing cases of iconic signification – to other sounds and phenomena.

First, when the case sounds are considered to embody similarities to other examples of sounds of foods, the fizzing sounds bear structural resemblance to the sound envelope emerging from, for example, the frying of a steak in a pan. Second, by comparison to sounds of non-food objects, the case sounds bear resemblance to the sounds from, for example, an intense applause by a large group of people. The sounds might also be likened to musical sounds. For example, the fizzing sounds appear similar to the short-lasting shimmering sounds produced by striking a slightly opened high hat (from a drum kit); and the composer H. C. Lumbye distinctively imitated the sound of the opening of a bottle of champagne at the beginning of his musical piece "Champagnegalop" – written to celebrate the second anniversary (in 1845) of Tivoli in Copenhagen. While these examples illustrate observations of the case sounds from the perspective of predominantly sonic anaphones, kinetic and tactile anaphones emerge when the sounds are compared to other sensory stimuli (for a similar approach, though focussed more on perception than on textual structure and including a focus on tastes, see the examination of crossmodal correspondences in Spence & Wang, 2015a). Already, a tactile anaphone has been exemplified by the observation of the sounds' "dryness," and further kinetic and tactile dimensions could arguably be observed by likening the sounds to the sensation of what is popularly described as "goosebumps." The progression of goosebumps – a possible part of the "sensory buzz" associated with AMSR (Lewis, 2019: 65) – thus seem to mirror the envelope of the fizzing sounds: A gradual increase followed by a flattening out of intensity in the form of myriads of (tingling and fizzing) microimpulses – perhaps tellingly, the sociologist Émile Durkheim offered the concept of "collective effervescence" (2001: 157ff) to account for the heightened arousal associated with certain forms of group-based interaction.

Causal listening and indexical signification

From the perspective of the causal mode of listening, the analysis addresses *from what the sounds emerge*. Correspondingly, by way of indexical signification – that is, a relationship between signifier and signified based on direct (causal or physical) connection (Chandler, 2007: 37) – sounds offer the listener an impression of the characteristics of the "object that has the sound" (Pasnau, 1999: 316) as well as aspects of the context of that object, that is, "an *interaction* of *materials* at a *location* in an *environment*" (Gaver, 1993: 6; italics in original). Arguably, the perspective thus concretely exemplifies how "[h]earing is a way of touching at a distance" (Schafer, 1977: 11). Often, reduced and casual listening emerge "in tandem" as the sounds' quality and structure form the basis for the assessment of, from what the sounds originate: "What leads us to deduce the sound's cause if not the characteristic form it takes?" (Chion, 1994: 32).

Gaver (1993) proposes a tripartite classification of object sounds based on differences between the attributes of objects and their interactions. Although Gaver does not focus on (digital) foods, the classification can help specify from what type of food a given sound originate. Firstly, "aerodynamic sound" is characterised by a "direct introduction and modification of atmospheric pressure differences of some source" (Gaver, 1993: 13), and the pop sound (at 0:15) exemplifies this type of sound. Aerodynamic sounds can appear continuously (e.g., the whistling sound from a tea kettle) or explosively, as illustrated by the pop sound. The pop sound markedly indicates – exemplifying how "[a]nything that focusses the attention is an index" (Peirce in Chandler, 2007: 42) – that a bottle of highly pressurised (carbonated) and hence "fresh" liquid has been opened. Also, the forcefulness of the pop indicates a quick release of energy and thus an unrestrained opening of the bottle. The pouring of the liquid, secondly, exemplifies a "liquid sound," which is generally "determined by the formation and change of resonant cavities in the surface of the liquid" (Gaver, 1993: 15). If it is indeed "easy to tell whether a liquid gurgling out of a bottle is water, or a thicker syrup or oil" (ibid.: 16; see also Spence & Wang, 2015d), then the splashing and fizzing sounds indicate the pouring of a carbonated liquid with a relatively low viscosity (or "thickness"). Thirdly, sounds from "vibrating solids" are exemplified by the interaction of the liquid sound with the particular container used. For example, the increased pitch of the fizzing sounds (observed previously) indicates that a relatively small and narrow container is being filled. The latter observation further illustrates the affinity of sounds "of" and "at" food mentioned at the beginning of the chapter: When we listen to "sounds of food," we do not, strictly speaking, merely hear the food substance. Rather, we hear the result of an interaction between the food substance and an assortment of materials (e.g., packaging and glassware) which in turn influences the quality and structure of the "sounds of food" in question. The case thus illustrates how sounds can provide information about the attributes (size and form) of objects as well as the interaction (type and force) between objects over time – illustrating how "[h]earing is particularly good at grasping 'the dynamics of things coming into being over time'" (van Leeuwen, 1999: 195).

In addition to the pop and pouring of carbonated liquid, the sounds inform us on what type of carbonated liquid we are (not) listening to. Although some soda and beer brands include opening mechanisms that produce a pop sound (Spence & Wang, 2015c: 2ff), the pop sound of the case is more forceful as the result of a higher level of carbon dioxide in a bottle of sparkling wine compared to a bottle of beer. Consequently, the pouring of beer into a glass would normally result in less vibrant fizzing sounds (Zhang & Xu, 2008). The example does however not provide information on a specific location in an environment, as there is no reverberation and cues of directionality (as observed previously).

While the sounds indicate, that we are listening to sparkling wine, the particular type is not clearly revealed by the sounds. Admittedly, different examples from within the generic category of sparkling wine does not always sound the same. For example, a study (Spratt, Lee, & Wilson, 2018) compared the sounds of Moët & Chandon Impérial with the sounds of Cook's California Sparkling Wine based on recordings gathered from a hydrophone immersed in filled glasses. A difference between the two recordings was found: The sounds of the Moët had a "slightly higher mean frequency, indicating that they [the bubbles] are slightly smaller overall than the Cook's bubbles" (Spratt, Lee, & Wilson, 2018: 67). Consequently, a higher mean pitch frequency might indicate smaller bubbles and thereby a better wine quality (bubble size is widely considered to correspond to the wine's quality; see, e.g., Vickers, 1991: 95). However, in the context of serving the champagne, the sizes of the bubbles vary according to, for example, the temperature, the surrounding atmospheric pressure, and the vessel into which the beverage is poured (Spratt et al., 2018). It is basically doubtful whether the technologically unequipped "naked ear" – situated on everyday listeners (and textual analysers) not actually immersed in the beverage – is able to perceive the recorded sound differences. Therefore, although the sounds might be explicitly associated with a particular type of sparkling wine through the specific digital media setting, there is no obvious indication from the structure of the sounds that they originate from champagne, let alone a particular brand of champagne.

Additionally, as already hinted at in the beginning at this chapter, we cannot be absolutely sure if the sounds originate from "the real thing," that is, champagne or some other type of sparkling wine. Based on the sounds' structural resemblance to other sounds (see above), the sounds might be "artificially" produced by use of, for example, synthesisers and/or produced from other objects. For example, apparently, the fizzing sounds can be imitated by "beans falling onto a plastic tray" (Knapton, 2015). The practise of so-called Foley artists in the context of mediated sound production further illustrates the point (for examples including foods, see Ament, 2014: 213ff). If the causal mode of listening is indeed "the most easily influenced and deceptive mode of listening" (Chion, 1994: 26), it holds particularly true of mediated settings, where the production of the sound does not necessarily coincide with the distribution of the sound and where the sound's source might not be visible.

Semantic listening and symbolic signification

From the perspective of semantic listening, the examination addresses *to what the sounds refer* beyond their resemblance to other sounds and the object and interaction of their origin. As was the case for the relationship between causal and reduced listening, semantic listening can overlap: "Obviously one can listen to a single sound sequence employing both the causal and semantic modes at ones" (Chion, 1994: 28). Semantic listening was exemplified at the beginning of the chapter where "champagne" was identified as a possible association: Whereas there is nothing from the structure of the sounds that distinctively establishes a relation to champagne, an association to champagne – and by extension the Champagne region, French nationality and/or specific social occasions (Guy, 2003) – might nevertheless be produced in the minds of specific listeners based on learning and convention.

Following the definition that the relationship between signifier and signified in the symbolic mode is "fundamentally *arbitrary* or purely *conventional*" (Chandler, 2007: 36; italics in original), the sounds might not qualify unambiguously as a symbol for champagne. As the sounds represent an actual feature (although not fully nor distinctively representative) of "champagne," the relationship between the sounds and "champagne" is partly motivated or "restrained" (ibid.: 38). Therefore, the association to "champagne" could be seen to include indexical aspects as well: The sounds appear "as a member of a class" that has come to "stand for the class," thus representing a subcase of a synecdoche, namely "*species for genus*" (Chandler, 2007: 133; italics in original). Similarly, "champagne" sometimes appears in daily discourse as a reference for the generic category of sparkling wine even though sparkling wines from only the region of Champagne are formally entitled (Sharp & Smith, 1991: 18). However, given that relationships between a signifier and a signified can be more or less motivated and conventional – for example, "[w]ithin each form signs also vary in their degree of conventionality" (ibid.: 38) – it seems fair to suggest that some degree of conventionality is involved if the sounds should come to stand for "champagne."

From a promotional perspective, the association between the sounds and champagne is produced through an extraordinary process. Rather than exemplifying the design of a distinctive and possibly rights protected variant of the generic sound in question – a strategy exemplified by the crunching sound of Kellogg's cornflakes (Lindström, 2005: 16f) and the engine roar of the Harley Davidson motorcycle (Sapherstein, 1998) – the association is historically established and cultivated through, for example, myths (Rokka, 2017) and promotional initiatives (e.g., advertisements, product placements, and sponsorships) that position the sounds in specific ways. Consequently, in addition to possibly indicate that champagne (and not any other type of sparkling wine) is made available for consumption, the sounds might produce associations to dimensions of human life, for example, certain feelings and social occasions. Such associations are dynamic and changes over time. Whereas champagne at the end of the eighteenth century was a "rare, luxury item [...] reserved for the wealthy" (Guy, 2003: 11) – and therefore a marker of social exclusivity and status – it has arguably since emerged as a more widespread ingredient of mass consumption. However, an association between champagne and the celebration of events "above the mundane" (ibid.: 32) seem to persist. For example, the largest champagne producers have continuously promoted their wine "primarily as a celebratory drink" (Walters, 2017: 16), and specific causes for celebration have included junctures or turning points in life, anniversaries (as exemplified above by the "Champagnegalop"), launching of ships and planes, and ceremonies at sporting events (Guy, 2003: 10ff). Again, the reference to celebration might not be based purely on convention. From the perspective of reduced listening, the reference could be seen to be somewhat motivated by the sounds' structural resemblance to similarly "exited" and "joyful" phenomena such as applause, "collective effervescence" and goose bumps. Moreover, from the perspective of causal listening, the relatively forceful pop of the bottle opening could seem to indicate a lifted mood where energy (and loud sounds) is allowed to be released without much restraint; by comparison, a "soft sigh"-opening sound could symbolise a more restrained and serious situation (e.g., at a wine tasting or during fine dining).

Key lessons and additional analytical perspectives

The preceding sections have demonstrated how to analyse food sounds when it comes to the "sounds of food" representing the core example of food sounds. The progression in three stages – and the analytical perspectives included at each stage – can be generalised to the analysis of other examples of "sounds of food." Figure 4.1 shows how the specific focus on a case of "sounds of food" (see the box marked by grey background) can be positioned within a broader framework of food sounds. The figure also indicates examples of how the examination can be widened to include other types of sound as well as other processes of food. For example, "sounds of food" relating to wine production could include the sounds from squeezing the grapes to evaluate their ripeness (see, e.g., Taber, 2005: 144f); and "sounds at food" could be heard in the storage environment: "The only rhythm here [the earth deep under Reims] is the slow drip of water, seeping through the sodden chalk" (Stelzer, 2017: 8). Although sound does not feature explicitly among, for example, the "six S's of wine tasting: seeing, swirling, smelling, sipping, spitting (or swallowing), and savoring" (Harrington, 2008: 23), slurping (with the purpose of airing the wine) might precede spitting thus exemplifying "sounds of food" consumption. To illustrate "sounds at food" consumption, imagine the sounds of clinking glasses and the crunching sounds from moving a bottle in and out of a wine cooler.

The object sounds "of" and "at" foods are relatively distinctive of each category of food process. For example, the sounds from unwrapping the foil of the champagne bottle are indicative of preparation more than of production or actual consumption.

Types of sound and relation to food \ Food processes		Production	Preparation	Consumption
Object sounds	Sounds of food	Crushing and picking of grapes	Pop and pouring of champagne	Slurping of champagne
	Sounds at food	Water dripping in a cave	Unwrapping the foil	Clinking of glasses and ice cubes in a bucket
	Sounds with food	Various object sounds neither 'for' nor 'at' food. For example, the sound of footsteps in a restaurant.		
Music		Various artists, genres, pieces, and types of distribution (e.g., live vs. recorded), etc. For example, the background music in a restaurant.		
Speech		Various modes of address, speakers, tone of voice, etc. For example, the sounds from conversations in a restaurant.		

FIGURE 4.1 Overview of food sounds exemplified by the case of sparkling wine.

The distinctiveness of the sounds' relation to food processes becomes clear(er) when compared to other types of foods. For example, the crispy and crunchy sounds of the consumption of lettuce and almonds (Vickers, 1991: 93ff) differs markedly from the sounds of slurping wine. By comparison, music and speech (and object sounds neither "of" nor "at" foods) have a much more "loosely coupled" relationship to the three food processes. Although, for example, correspondences between certain types of music and certain styles of wine have been reported (see Spence & Wang, 2015a), there is no distinct relation between "sounds with food" and the three food processes (for textual examination of speech sounds' mode of address and significance, see Stigel, 2001; Tagg, 2013: 343ff).

A further broadening of the analytical interest concerns the type or genre of digital texts, where some (sounds from) processes of food are more pronounced than others. For example, instructional "how-to-cock" videos would include mostly sounds of preparation, restaurant settings in films would include mainly sounds of consumption, and documentaries on specific types of foods might include a relatively large proportion of sounds of production. Other digital texts might include all three processes, for example, in the context of advertising. Although the introductory example illustrates how object sounds alone can constitute a case of digital foods (granted that the accompanying visuals on YouTube is not a significant part of the mediated text), these examples furthermore illustrate, that digital texts often include more than "just" sounds of food. The analysis of food sounds should then include an examination of the relations between different types of sound and the relation between sounds and the visuals (for a general introduction to textual analysis of audiovisual digital media, see Chion, 1994; Graakjær, 2015; and for a case analysis of sounds of food, see Graakjær, 2021). Moreover, the analysis could include a critical examination of how relations between sounds and foods might contribute to the idealisation of certain foods and to the staging and (re)production of power imbalances and ideology (for a case analysis of ideological implications of sounds for food, see Graakjær, 2019). For example, as they hold the potential to symbolise celebratory occasions and a certain way of life, the sounds of (and at) sparkling wine might function as an auditory underlining of who (in terms of, e.g., gender, ethnicity, and social status) can (not) enjoy privileged access to those occasions and ways of life. This would include an analysis of who is seen or implied to produce and listen to the sounds. While it is beyond the scope of this chapter to discuss in further detail how such a contextual examination could be performed, the chapter has indicated the significance of food sounds: Sounds can influence our perception of food types and qualities, and sounds can indicate the social and physical settings in which various processes of food handling emerge. Hopefully, the chapter's three-staged model for the examination of food sounds can serve as an inspiration for textually oriented digital food studies.

Dos and Don'ts

Do accept, that an analysis of digital food sounds should be based on "repeated hearings" of the given sound sequence.

Do try to "test" the distinctiveness and significance of food sounds in the digital media setting by performing so-called commutation tests, that is, to observe the consequence of substituting, reordering, adding, or deleting specific auditory elements while holding all other elements constant. The procedure can prove fruitful even when performed as a "mind game" (for more on commutation tests, see Graakjær, 2015: 104–109).

Don't rush to adopt the casual and semantic mode of listening. A proper analysis of digital food sounds includes analytical observations based on reduced listening as well.

Don't expect the analysis to deliver nothing else/less/more than *potentials* of meanings. Actual meanings (based on reports from actual listeners) could be examined from the perspective of reception studies.

Now you try

Choose another food example (e.g., cereal, fruit, chocolate-coated ice-cream) from a digital text (e.g., a commercial or an AMSR-video) and examine the sounds by following the demonstrated three stage analytical process: what do the sounds sound like, from what do they originate, and to what can they be considered to refer symbolically? Also, try to identify and discuss the sounds in terms of their status seen from the perspectives of food processes and relations to foods – that is, fill out the boxes of Figure 4.1 with reference to the chosen food example.

Questions for reflection

Based on a selection of three different, comparable digital media texts including the sounds of champagne (e.g., three comparable scenes from films or three websites for champagne brands), examine the following questions: How do these sounds of champagne compare to the sounds of the chapter's case and how do they mutually compare?

What are the possibly distinctive features and meanings of food sounds in digital texts compared to other types of sound (e.g., the sounds of and at a car)?

What is the possibly distinctive contribution of digital food sounds compared to digital food visuals?

Ethical considerations

In particular, three ethical issues must be considered when performing an analysis of digital food sounds from a textual perspective.

Accessing the analysed text. If you link to the analysed text(s), you need to make sure, that the link target (i.e., the place where the text can be accessed for "repeated hearings") is public, legal, and in fact accessible – the present paper exemplifies how these criteria can be met.

Representing and redistributing the analysed text. If you include a representation of the sounds in your publication (e.g., a video clip integrated in the publication), you need to make sure to comply with copyright law and fair use – the present paper does not exemplify a representation and redistribution of the analysed text.

Person data. If the text includes information that identifies or makes identifiable specific persons, you need to make sure to meet the standards of the GDPR – the present paper does not exemplify the inclusion of person data relevant for the GDPR.

Further readings

Graakjær, N. J. & Bonde, A. (2018). Non-musical sound branding – a conceptualization and research overview. *European Journal of Marketing*, 52(7/8), 1505–1525.

Spence, C. & Wang, Q. (2015). Wine and music (II): Can you taste the music? Modulating the experience of wine through music and sound. *Flavour*, 4(33), 1–14.

van Leeuwen, T. (1999). *Speech, music, sound*. London: Macmillan Education.

References

Ament, V. T. (2014). *The Foley grail: The art of performing sound for film, games, and animation.* Abingdon: Focal Press.

Capeller, I. (2018). Sounds, signs and hearing: Towards a semiotics of the audible field. *Athens Journal of Philology*, 5(1), 45–60.

Chandler, D. (2007). *Semiotics. The basics.* Abingdon: Routledge.

Chion, M. (1994). *Audio-vision – Sound on screen.* New York: Columbia University Press.

Danesi, M. (2007). *The quest for meaning. A guide to semiotic theory and practice.* Toronto: University of Toronto Press.

Durkheim, E. (2001). *The elementary forms of religious life.* Oxford: Oxford University Press (original French edition, 1912).

Epidemic ASMR (2017, May 16). *Champagne Bottle Open Sound.* Retrieved from https://www.youtube.com/watch?v=mJxAtXfOGts

Gaver, W. (1993). What in the world do we hear? An ecological approach to auditory source perception. *Ecological Psychology*, 5(1), 1–29.

Graakjær, N. J. (2015). *Analyzing music in advertising: Television commercials and consumer choice.* New York: Routledge.

Graakjær, N. J. (2019). Sounding out *i'm lovin' it* – a multimodal discourse analysis of the sonic logo in commercials for McDonald's 2003–2018. *Critical Discourse Studies*, 16(5), 569–582.

Graakjær, N. J. (2021). Sounds of Coca-Cola – On "cola-nization" of sound and music. In J. Deaville, R. Rodman, & S.-L. Tan (eds.), *Oxford handbook for music and advertising* (pp. 397–413). Oxford: Oxford University Press.

Graakjær, N. J. & Bonde, A. (2018). Non-musical sound branding – A conceptualization and research overview. *European Journal of Marketing*, 52(7/8), 1505–1525.

Guy, K. (2003). *When champagne became French: Wine and the making of a national identity.* Baltimore: The John Hopkins University Press.

Harrington, R. J. (2008). *Food and wine pairing. A sensory experience.* Hoboken: John Wiley & Sons.

Knapton, S. (2015, 1 May). What does wine sound like? You're about to find out. *The Telegraph.* Retrieved from https://www.telegraph.co.uk/news/science/science-news/11577381/What-does-wine-sound-like-Youre-about-to-find-out.html

Lewis, T. (2019). *Digital food. From paddock to platform.* London: Bloomsbury Academic.

Lindström, M. (2005). *Brand Sense: How to build powerful brands through touch, taste, smell, sight & sound.* New York, NY: Free Press.

Pasnau, R. (1999). What is sound?. *The Philosophical Quarterly*, 49(196), 309–324.

Rokka, J. (2017). Champagne: Marketplace icon. *Consumption Markets & Culture*, 20(3), 275–283.

Sapherstein, M. P. (1998). The trademark registrability of the Harley Davidson roar: A multimedia analysis. *Boston College Intellectual Property & Technology Forum.* Retrieved from http://bciptf.org/ wp-content/uploads/2011/07/48-THE-TRADEMARK-REGISTRABILITY-OF-THE-HARLEY.pdf

Schafer, M. (1977). *The soundscape. Our sonic environment and the tuning of the world.* Rochester: Destiny Books.

Schifferstein, H. (2016). The roles of the senses in different stages of consumers' interactions with food products. In B. Piqueras-Fiszman & C. Spence (eds.), *Multisensory flavour perception. From fundamental neuroscience through to the marketplace* (pp. 297–312). Amsterdam: Woodhead Publishing.

Sharp, A. & Smith, J. (1991). Champagne's sparkling success. *International Marketing Review*, 8(4), 13–19.

Spence, C., Carvalho, F. R., Velasco, C., & Wang, Q. J. (2019). Extrinsic auditory contributions to food perception & consumer behaviour: An interdisciplinary review. *Multisenory Research*, 32(4–5), 275–318.

Spence, C. & Wang, Q. (2015a). Wine and music (I): On the crossmodal matching of wine and music. *Flavour*, 4(34), 1–14.

Spence, C. & Wang, Q. (2015b). Wine and music (II): Can you taste the music? Modulating the experience of wine through music and sound. *Flavour*, 4(33), 1–14.

Spence, C. & Wang, Q. (2015c). Wine and music (III): So what if music influences the taste of wine? *Flavour*, 4(36), 1–15.

Spence, C. & Wang, Q. (2015d). Sensory expectations elicited by the sounds of opening the packaging and pouring a beverage. *Flavour*, 4(35), 1–11.

Spratt, K., Lee, K., & Wilson, P. (2018). Champagne acoustics. *Physics Today*, 71(8), 66–67.

Stelzer, T. (2017). *The champagne guide 2018–2019: The definitive guide to champagne*. Richmond: Hardie Grant Books.

Stigel, J. (2001). The aesthetics of Danish tv-spot-commercials. A study of Danish TV-commercials in the 1990'ies. In F. Hansen & L.Y. Hansen (eds.), *Advertising research in the Nordic countries* (pp. 327–350). Frederiksberg: Samfundslitteratur.

Taber, G. (2005). *Judgment of Paris*. New York: Scribner.

Tagg, P. (2013). *Music's meaning. A modern musicology for non-musos*. New York: MMMSP.

van Leeuwen, T. (1999). *Speech, music, sound*. London: Macmillan Education.

Vickers, Z. (1991). Sound perception and food quality. *Food Quality*, 14(1), 87–96.

Walters, R. (2017). *Bursting bubbles. A secret history of champagne and the rise of the great growers*. Abbotsford Vic: Quiller.

Wang, Q. J., Mielby, L. A., Junge, J. Y., Bertelsen, A. S., Kidmose, U., Spence, C., & Byrne, D.V. (2019). The role of intrinsic and extrinsic sensory factors in sweetness perception of food and beverages: A review. *Foods*, 8(6), 1–26.

Zhang, Y. & Xu, Z. (2008). "Fizzics" of bubble growth in beer and champagne. *Elements*, 4, 47–49.

5

HOW TO APPROACH FOOD TEXTS ON YOUTUBE

Thomas Mosebo Simonsen and Stinne Gunder Strøm Krogager

In this article, we discuss and exemplify how to work with YouTube as a vast source of food-related content. We illustrate how to identify different modes of audiovisual representations of food and, simultaneously, we reflect upon a methodological approach to analysing and identifying various food videos on YouTube. Handling large samples of videos raises the fundamental methodological question on sampling with regard to selecting and identifying relevant and representative data. In the following, we introduce the very latest research on YouTube food, after that, we outline the different approaches to working with YouTube, and then we exemplify how to approach YouTube in a small case study. Summing up, we reflect on the opportunities and challenges when working with YouTube.

YouTube composes a large and diverse compilation of audiovisual material on food and cooking, and food is the fastest growing genre on the platform. Advice videos related to cooking are in the top 10 most popular "how to" searches on YouTube, according to Google (Lewis, 2018). Correspondingly, food and cooking

TABLE 5.1 YouTube facts

- YouTube is the world's largest online streaming site and the second most visited website globally.
- YouTube was established in 2005 and bought by Google Inc. in 2006.
- More than 2 billion users upload 500 hours of content every minute.
- All types of content are available – music videos, movies, television content, UGC, etc. - and over 70% of views are consumed on mobile devices.
- It has a widespread audience existing in 80 different languages and it reaches more 18–34 year olds that any TV network.

YouTube.com – facts, 2020

on YouTube is a rather well-researched area within the disciplines of consumer and marketing studies, public health, and medicine, especially paediatrics and obesity (e.g., Yoo & Kim, 2012; Worsley et al., 2014; Pereira et al., 2019). Yet, scholarly attention from the perspective of media, communication, and cultural studies has, up to now, been very scarce. Recently however, prominent works with different approaches and emphases on YouTube and food have been issued.

In the recent book, *Digital Food: From Paddock to Platform* (Lewis, 2020), Lewis examines the intersection between food and the digital realm and thus, how digital devices and digital content on food saturate everyday life. The book offers several perspectives of the field, and many of these delve specifically into YouTube. In one chapter, Lewis considers the notion of "ordinary expertise" performed in YouTube "how to" cooking videos and also reflects on the rise of professionalised cookery channels and brand cooks and, consequently, the growing monetisation of YouTube and similar platforms. In another chapter, Lewis examines whether the masculinities that are represented in hipster "bros" cooking on YouTube transform the politics of domestic labour. Thus, the book provides knowledge on YouTube, food and cooking from different analytical perspectives.

In the equally recent edited volume *Digital Food Cultures* (Lupton & Feldman, 2020), Lupton and Feldman include two case study pieces on YouTube: Lupton's own contribution focuses on a genre of YouTube videos that she calls "carnivalesque." She scrutinises two examples that portray excessive food consumption. These videos create fascination through the transgressive and grotesque indulgence of large amounts of calorie-heavy foods and, through this excessive preparation and consumption, sexualised stereotypes of hyper-femininities and hyper-masculinities are being reproduced (Lupton, 2020b). Braun and Carruthers' contribution analyses different genres of vlogs produced by practicing vegans and conclude that the vloggers are predominantly young, white, able-bodied women with lean bodies that depict the vegan diet in a food porn kind of way (Braun and Carruthers, 2020). Both contributions in this volume illustrate the celebration of the multisensory desires of food, however, in different degrees and in both contributions, gendered norms and stereotypes come into play.

Another recent work about YouTube that focuses on gendered norms and performances is Contois' commentary on the YouTube series *Hot Ones* (Contois, 2018). Contois investigates how celebrities, mainly male, navigate gendered performances during a conversational interview while consuming chicken wings and hot sauce. A recurring theme in the show is the gastric distress that may follow the intake of spicy food and talking about digestion topics and indulging in fried chicken wings with spicy sauce is far from culturally constructed feminine conventions of bodily control and a restrained appetite.

In an article from 2017, Forchtner and Tominc study veganism. Building on Barthes (1961), Douglas (1966), and Fischler (1988), among others, they argue that food consumption is a profoundly emblematic topic that defines individual and collective identities. Accordingly, veganism and vegetarianism are linked to peace-loving and left-leaning ideologies; however, through analyses of videos and interviews,

Forchtner and Tominc (2017) demonstrate how a group of neo-Nazis (*Balaclava kitchen*) adopts a vegan diet in its YouTube cooking videos and also, how traditional extreme right hyper-masculinity, usually closely linked to eating meat, is being performed differently in this group. In particular, this article accentuates how cooking and food consumption are areas that profoundly intertwine with gendered norms and politics.

Gender norms, construction, and performance are recurring motifs in all contributions. Besides these, politics is a significant theme and also emphasises a democratic perspective on YouTube; the platform has empowered anyone with internet access and sufficient skills to partake in the dispersal and sharing of ordinary, amateur expertise. "Ordinary" people offer guidance and provide know-how on food and cooking-related topics and thus, the lines between professionals, celebrities, and amateurs are increasingly indistinguishable. These amateur social media influencers are one of the most distinct categories on YouTube (Lewis, 2018) and they play an imperative part in generating attention for food cultures and trends (Lupton, 2020a). Methodologically, all contributions draw on qualitative methods from the traditions of media and cultural studies, applying different types of textual analyses. This small-scale approach prevails in academic studies on YouTube, from the perspective of the humanities and social sciences, and we will elaborate on this in the following section.

Approaches to working with YouTube: Qualitative and small-scale studies versus quantitative and large-scale platform studies

The methodologies applied when researchers investigate YouTube can be divided into two tendencies: qualitative, small-scale studies and quantitative, large-scale platform studies. The first is related to an agency-driven analysis, with a primary interest in social interaction and agency, for example, with regard to parenting or learning (Lee & Letho, 2012; Quennerstedt et al., 2013; Burroughs, 2017), social interaction is specifically investigated within the audiovisual content (e.g., Smith, 2014; Cunningham et al., 2016; Berryman & Kavka, 2018) as well as the investigation of YouTube's production culture (Burgess & Green, 2009/2018; Strangelove, 2010). This research tends to employ qualitative approaches, such as interviews (Chen, 2016; Gibson, 2016); or a small-scale sample approach (Lovelock, 2017). These contributors also share the aforementioned democratic discourse and a digital optimism, describing the emergence of user empowerment. Originally founded in the politics of cultural studies, and brought up to date by, for example, Bruns (2008), Jenkins (2006) and, most noticeably with regard to YouTube, by Burgess and Green (2009/2018).

An early contributor was Lange (2007). Via an ethnographic approach, combining semi-structured interviews with both online and offline participants, Lange investigated how YouTube affected social interaction. Burgess and Green's *YouTube – Online Video and Participatory Culture* (Burgess & Green, 2009/2018) is the first monograph and still the most cited investigation of YouTube. Based on a sample of YouTube content from 2007 (updated in 2018), the book's principal arguments are relevant in terms of understanding the mechanisms of the media industry in between commercial and user-generated content as participatory culture. Strangelove (2010) also contributed with an ethnographic investigation of YouTube's participatory culture based on personal interaction with the platform. A popular field of study within the interactive tendency is the types of content, such as the vlog, investigating audiovisual self-presentations, and self-branding focusing on intimacy and authenticity, as well as celebrity status (e.g., Berryman & Kavka, 2018; García-Rapp, 2017; Cunningham et al., 2016; Gibson, 2016; Smith, 2014). Several of these studies ground their approach in case studies of what we may regard as celebrity representatives of YouTube cultures, such as PewDiePie (gaming) (e.g., Cunningham et al., 2016) or Zoella (beauty and fashion) (e.g., Berryman & Kavka, 2018). Another heritage from cultural studies can be noticed in identity-driven investigations of gender (Molyneaux et al., 2008), race (Chun, 2013; Guo & Harlow, 2014) as well as queer studies (Lovelock, 2017). Lovelock (2017) performs a cultural textual analysis of lesbian, gay and bisexual identities and how digital media may influence emblematised social practices.

The focus is on mediated social interaction or audiovisual self-representation, and on the content rather than perception. Consequently, the applied approach in the publications within this field is most frequently case-driven. In fact, the case-driven analysis seems to have become the most widespread analysis within the field of communication, media studies, and journalism.

In contrast to Lange's (2007) in-depth and small-scale analysis and Strangelove's (2010) ethnographic orientated approach, Cha et al. (2007) provide one of the first large-scale, empirical investigations of YouTube, in which they discuss "the popularity life cycle of videos" (p. 1). Their contribution demonstrates the second tendency, which includes YouTube's materiality and its status as a platform. Drawing on a social network analysis (SNA), Cha et al. (2007) demonstrate that the popular content of YouTube is controlled by information filtering and algorithms, drawing attention to the structural mechanisms of YouTube. An SNA, as Cha et al. (2007) perform, describes the overall lines of content circulation, but does not provide more detailed distinctions of the content. However, a relevant contribution of their study is the awareness of the media platform's structural impact on the opportunities and constraints of gathering data and information, such as the role of algorithms and application programming interface (API), the programme that allows third party applications to embed YouTube videos into their platforms (van Dijck, 2013). The "platform" tendency frequently focuses on YouTube's interface and organisational features and how YouTube as a media platform may affect the communication and organisation of content (Chatzopoulou et al., 2010; Pinto et al., 2013), also involving examinations of more specific parts of the platforms, such as the effects of algorithms (Rieder et al., 2018;) or technical affordances (Postigo, 2016). The applied methodologies are frequently related to large-scale quantitative investigations, such as content analysis (Thorson et al., 2013; Whitaker et al., 2014), large-scale sampling (Halpern & Gibbs, 2012; Bärtl, 2018), SNA or platform studies (Bishop, 2018). Cha et al. (2007) introduced the integration of third-party apps (via API) as a methodological tool to access data, which has developed into a widespread approach gathering empirical content or metadata via API, for example, also applied by Bärtl (2018).

Additionally, the platform perspective raises an awareness of the increasing commercialisation of YouTube, creating an analytical counterpart and social critique of the digital optimism and user empowerment encompassing earlier studies of YouTube. Critical studies also combine platform analysis and critical media ecology analysis with, for example, social theory and actor–network theory (e.g., van Dijck, 2013; Fuchs, 2012; Vonderau, 2016). Lobato (2016) provides useful insights into the application of a media ecological approach identifying YouTube's cultural economic logic from an institutional perspective combined with specific selective cases, suggesting a methodological blend between a macrolevel institutional analysis and a micro-level case study.

These two research tendencies (small-scale and qualitative versus large-scale and quantitative – of which most of the research on food and YouTube is small-scale and qualitative, including the analytical case in this chapter) are relevant to have in mind during the process of investigating social cultural phenomena when either applying YouTube as a methodological tool or investigating YouTube as a field of research in itself. The interactive perspective provides relevance in terms of investigating social interaction as well as the content itself, for example, based on a case study analysis that may also draw on existing theories on social phenomena and interaction. As such, YouTube content is often analysed in comparison with already existing forms of audiovisuality or theories of social interaction, thus following a contextualised approach as proposed by Markham and Baym (2009: xv) in terms of involving aspects that both predate current media platforms, but are simultaneously adapted for it: "the commitment to making sense of the new by understanding their research processes' and object's continuity with the past." Simultaneously, we argue for an emphasis on the significance of analytical awareness to distinctive media platforms in terms of features and affordances, as well as the influence of contextual media ecological aspects and the increasing commercialisation of ordinary user cultures. When more specifically focusing on the implementation of affordances, the analytical focus point may usefully be built on existing research on affordances that furthermore acknowledge the multifaceted nature and relational understanding of affordances (Evans et al., 2016), when investigating how new technologies might affect digital food representation or interaction in a less deterministic approach.

Media platforms in flux

Perhaps the biggest challenge when studying YouTube and other media platforms is their constant changeability (Cf. Kjær's chapter on brand websites). Media platforms are in flux, repetitively evolving. Technological changes and user-driven innovations are defining media platforms as dynamic systems (Ellison & boyd, 2013). The researcher therefore needs to consider the fundamental state of instability of digital platforms. If we look at YouTube, although its principal interface is still intact, it has continuously changed formats, features, and forms of user engagement. YouTube has also changed its parameters for browsing, engaging and uploading several times. A study of YouTube is therefore always a study of the platform at a specific time period. Consequently, a diachronic analysis, examining deviations over time, requires a contextual approach that takes the dynamism of media platforms into consideration, since metadata, annotations, and channel interface may have altered. For example, specific parameters of popularity are continuously changing. Previously, YouTube videos could be favoured or rated instead of liked, just like YouTube provided specific charts, as used by Burgess and Green (2009/2018).

As specified by Cha et al. (2007) and further investigated by van Dijck (2009, 2013), the architectonical infrastructure of YouTube automatically promotes already popular content. Accordingly, many samples or case studies are selected within the most popular content (e.g., Burgess & Green, 2009/2018; Welbourne & Grant, 2016), also including celebrity studies on YouTube (Smith, 2014; García-Rapp, 2017). This may also be explained by the fact that YouTube, via its interface and algorithms, automatically provides access to popularity parameters, such as "recommendations" or the "trending list" that has replaced the previous chart lists. Data on YouTube can also be gathered by using YouTube's search engines. Molyneaux et al. (2008) and Kim et al. (2010) based their investigations on gender and smoking fetish videos on the YouTube search engine. Both investigations assumed that all videos were tagged in correspondence with the actual content. However, this is not always the case on YouTube, as tags are often used strategically in order to increase the video's search engine optimisation (SEO). Furthermore, search-based samples are not representative of all content. On many digital media platforms, there are no clear institutional regulation or conventional understanding of the specific word defining the content (via tagging). With the emergence of user-generated content, new meanings and words related to novel types of content continuously emerges. An example is "Mukbangs," that has no conventional or taxonomic meaning, but can be referred to as a "folksonomy" (Sinclair & Cardew-Hall, 2008). An idiosyncratic and less formal tag or keywords are created within specific social networks, thus depending on social and contextual meaning. According to Golder and Huberman (2006: 199), folksonomies are neither hierarchical nor exclusive categories and may appear in infinite combinations. The use of folksonomies may also be platform specific ('selfie' is for example, associated with Instagram) or geographically distinctive. "Mukbangs" are also spelled "muk-bang" or "meokbang," and a keyword search on YouTube generates three different search results, with "meokbang" displaying a majority of Korean-spoken videos, while "Mukbang" primarily results in English-spoken videos. Consequently, isolating a representative sample becomes more difficult. A contextual approach may enhance the process, requiring that the researcher needs to gather information and observe the interaction around the content within the social network from which the sample or case is isolated and gathered.

Moreover, similar to charts, results gathered from a search-based investigation are most likely governed by parameters of popularity. The selection of content based on search algorithms corresponds to how Snelson (2013: 5) characterises a sample-method on YouTube as a "top-down sampling" approach founded in the specific media platform's distinctive modes of organising content. YouTube's search engine is built upon Google's PageRank algorithms, promoting already popular content. Consequently, results based on a keyword search are not necessarily more accurate in correspondence with the actual content, but are possibly the most promoted or popular. This implies that researchers must distinguish between relevance and popularity. In their investigation of "pelvic floor muscle exercises" on YouTube, Stephen and Cumming (2010) apply a snowball approach. They used the search results to locate other suggested videos based on the previous search, in order to mirror actual viewer behaviour. Their approach is relevant to understand search patterns, but not necessarily more representative, as it excludes all the potentially relevant content not generated by YouTube's recommendation algorithm.

Accordingly, the task of compiling a representative sample is a challenge on YouTube with the unprecedented amounts of content and a constantly evolving platform always characterising a sample as a snapshot of YouTube at a specific time. A sample is most likely dependent on algorithms and popularity parameters, but it can be enforced by an extension of the contextual perspective of the communicative situation that demands a more active and interpretative role of the researcher. This may also involve research via social networks, involving users, performing a more collaborative approach (Havalais, 2013). With regard to food studies, we may navigate via publications and reflections on existing food channels to isolate a contextual understanding of the pragmatic meaning of food channels, how they are perceived or interacted with to isolate a sample based on relevance.

All the data

As Lovink stated back in 2008: "we no longer watch films or TV, we watch databases" (Lovink, 2008: 9). Content is not only content but also simultaneously data. With the gathering of data via digital media, we can identity a methodological turn that creates an unpreceded ability to collect large amounts of data based on automated computer-assisted analysis. This is most noticeable with the emergence of Big Data (Schönberger & Cukier, 2013). Computers allow researchers to include entire populations, thereby eliminating the process of sampling in order to focus on and predict large-scale patterns of communications. A direct impact of big data is datafication or the monetisation of data, in which user and content information are rendered into data and thereby given value. Datafication has a direct impact on researchers' limited access to data and may result in an increasing gap between research institutions and the media industry (Bruns, 2013). Although YouTube provides a large scale of accessibility in terms of access to all content, insights into algorithms, charts, and demographic user data are becoming increasingly more difficult to access. Platform corporates, such as Google or Facebook, are reluctant to provide data access as a direct consequence of the monetisation of data. Therefore, we notice a movement towards third-party agents in terms of statistic user data, using API.

When Chatzopoulou et al. (2010) investigated popularity parameters on YouTube, they gathered a sample of more than 37 million videos via API that helped organise and collect data but they were limited to the search patterns and algorithms of the constraints of third-party apps. API is a principal tool for global-tech companies, such as Facebook's and Google's ability to gather information and data from its users. An API enables content embedding, meaning that third-party applications may use and embed YouTube videos in their own platform. For researchers, this means not only access to information but also access that is potentially biased and subjective in terms of relevance and the commercial interests of the app provider.

Large-scale investigations are primarily interested in predictions and mapping of, for example, search patterns. As stressed by Arthurs et al. (2018), API requires knowledge of programming and statistics which, in many cases, makes the examiner dependent on other academic contributors, emphasising a necessary turn towards collaborative investigations across different fields of research. We may also apply third party apps to scrap and organise (without programming skills) small-scale samples based on themes, tags, or creators, and use them as contextual information relating to the specific platform. This includes the Google Chrome plugin vidIQ, which provides useful metadata about the video in terms of interaction, visibility on other media platforms and the specific tags connected to the video. Another API-driven app is Socialblade.com which provides access to YouTube's most popular channels, videos and creators in terms of views and subscribers. However, plugins like vidIQ or Socialblade are similar to YouTube's or Facebook's data limitations, increasingly defined as barriers to entry, where payment is required.

The case of *Tasty*

In the following section, we shall provide an example of a qualitative-driven case study that has been selected via the above-mentioned search patterns. Starting from a contextual investigation of food channels, the food channel *Tasty* is perhaps the most quoted and well-known food channel outside YouTube. It is also the most popular food channel on YouTube in terms of subscriptions and view counts. It has more than 17.2 million subscribers and 3,500 videos uploaded since 2016. Of course, to select *Tasty* as a case example following YouTube's search function results in a sample dependent on YouTube's algorithms and popularity parameters. Nonetheless, we have chosen *Tasty* as a case as it provides us with many different methodological options and analytical directions, making it a useful and relevant example of how to investigate YouTube food. The case is thus selected not only in terms of popularity but also because of its relevance, exemplified by how the channel produces variant types of food videos.

With more than 3,500 videos, the researcher is left with a lot of content. In order to gather or narrow down the content, we initially performed a quantitative registration in terms of a content analysis (Krippendorff, 2013) creating a sample to systematically register and count different variables to, for example, identify specific patterns of communications or food representations. A content analysis can often be sufficient in itself, but it can also be combined or replaced with a selection of a smaller, qualitative sample that involves a more detailed interpretative focus on the specific content. We use the available sorting parameters provided by YouTube, that is either view-counts creating a sample based on popularity or a sample based on the time of upload. The latter is more random and we can simultaneously investigate the most recent uploaded videos in order to get an up to date sample of the channel. The sample could also be gathered in terms of keywords, titles or thumbnails turning attention towards a contextual analysis that may help us register specific types of food, visualisations of food, or to provide us with a tool to identity popular food topics. Accordingly, the process of isolating the sample is framed by analytical relevance and narrowing.

Tasty is part of the multichannel network, Buzzfeed, a large entertainment and news provider that uses several media platforms and distributes content on several YouTube channels, among these, the food channel *Tasty*. The contextual approach to *Tasty* is to consider it an example of the hybrid economy ofYouTube (Arthurs et al., 2018; Burgess & Green, 2009/2018;) functioning as a corporate channel but simultaneously applying communicative and aesthetic features related to an everyday aesthetic (Vernallis, 2013). Accordingly, *Tasty* furthermore provides interesting examples of how food is aestheticised and communicated within a media-ecological contextual perspective on the commercial role of corporate content food, for example, with regard to locations, audiovisual presentations of food or contextual communication in terms of metadata, annotations, or paratextual readings between audiovisual content, user comments, and promotional texts below the videos.

Tasty was originally produced for Facebook. It would therefore also be relevant to examine how the content differs whether it is distributed on YouTube or Facebook with regard for format and user face. This furthermore involves a perspective on how affordances might lead to platform specific features and social actions, for example, on the role of the Facebook constraint in terms of videos auto playing silently in the news feed, as well as interactive differences between the two platforms, elaborating on how food might be represented differently on the two platforms.

When investigating the contextual influence of media platforms in terms of food presentations, it is therefore relevant to motivate the choice of platform, to perform a comparative analysis between the two platforms, and how this may influence the presentation of food.

Tasty provides several different types of content that call for a typological analysis (Simonsen, 2011). We apply theoretical knowledge on existing formats to identify the different modes of addressing the audience. For example, the "how to" format seems to be the most frequent type of video on the *Tasty* channel. It resembles a didactic, instructional form of communication, analogous to audiovisual modes of representation within the documentary and news format, from which rhetorical strategies demonstrate how presentations of food provide a learning-perspective as a rhetorical strategy to include its audience, however differing from traditional instructional content (e.g., on television) in terms of aesthetics and affective strategies. Another popular format is the vlog, highlighting the personal presenter of the food rather than the food itself. With regard to the vlog, the interaction between the host and the audience exemplifies a parasocial interaction (Horton & Wohl, 1956/2006) or personal appearance and interaction (e.g., front–back-region perspectives, Goffman, 1959/1990). The communication between creators and audience can also be regarded as an expression of sociability, also including the written user comments below the video.

Finally, our analytical approach based on relevance also includes a thematic investigation of food representations. For instance, what types of food are being presented and how it is specifically presented? Summing up, this is of course only a case example in headlines, but it helps to illustrate the need for a contextual approach that begins outside the YouTube platform. It also acknowledges the commercial context and influence of algorithms and API when selecting the case. This is nuanced by investigating the contextual information in terms of metadata and user reactions. We must furthermore accept the fact that one case or sample cannot provide us with any general information, but investigating a random sample of recent content allows us to present an isolated insight into the dynamic platform mechanisms of YouTube.

Dos and Don'ts

As we outline above, working with YouTube entails several unknown parameters with regard to search and algorithm patterns. Hence, using YouTube as a point of departure for analysing digital food entails a meticulous focus on the sampling or selection of texts/cases/channels. Whether it is a small-scale in-depth analysis of one single food video or a large-scale quantitative content analysis of many texts, argumentation for the selection of text(s) is vital for the trustworthiness of the research (Bryman, 2016: 384–386).

Do make sure that your arguments for choosing the specific text is clear and theoretically well-founded.

Do elaborate on all aspects of the methodological design, for example, sampling of texts (how have you selected the texts from YouTube?) and methods of data analysis (are you doing a small-scale or large-scale analysis?).

Do pay attention to the scope of your research design: what does your analysis allow you to conclude? For example, an analysis of *Tasty* (as our example above) does not allow us to say anything in general about food channels on YouTube.

Don't skip methodological argumentations even when they seem over detailed. These arguments secure the reliability of your research.

Don't attempt to include everything. The all-encompassing analysis is a misconception. Rather focus on details and subjects relevant for your research question.

Now you try

... to outline a research design on YouTube food. First, frame a research question (RQ) that you find motivating. The RQ will, to some extent, determine which methods you need to use (Bryman, 2016: 403): does the RQ prepare the ground for a qualitative in-depth approach or a broader quantitative approach? Whichever method is appropriate, then move on to identifying and describing what type(s) of text(s) you are dealing with and finally, what kind(s) of analytical approach(es) you could potentially use to examine the text(s) thoroughly and will eventually answer your RQ.

Questions for reflection

How can we approach and select YouTube materials for analysis given the fact that we will never fully know the underlying algorithms of the platform (or any other digital (social) media platform)?

How can we respond to criticisms that an analysis of *one* YouTube video or one YouTube channel does not inform us about anything on YouTube food in general?

What kinds of ethical considerations do we need to have in mind when working with public material from, for example, YouTube?

Ethical considerations

YouTube is an open and public platform. That is, anyone with internet can access it. Also, the aim of a YouTube channel is most often to get as many views, likes, comments, and followers as possible. However, that does not mean that ethical concerns are not relevant. When dealing with online content, it is always important to reflect on issues of privacy. With regard to the example of *Tasty*, we would draw on the material that the channel uploads, without a formalised consent. However, if we wanted to include the comments from followers, we would anonymise the material.

Further readings

Arthurs, J., Drakopoulou, S., & Gandini, A. (2018). Researching YouTube. *Convergence: The International Journal of Research into New Media Technologies*, 24(1): 3–15.

Braun, V., & Carruthers, S. (2020). Working at self and wellness: A critical analysis of vegan vlogs. In D. Lupton & Z. Feldman (eds.), *Digital Food Cultures* (pp. 82–96). London: Routledge.

Forchtner, B., & Tominc, A. (2017). Kalashnikov and cooking-spoon: Neo-Nazism, veganism and a lifestyle cooking show on YouTube. *Food, Culture & Society*, 20(3): 415–441.

Lupton, D. (2020b) Carnivalesque food videos: Excess, gender and affect on YouTube. In D. Lupton & Z. Feldman (eds.), *Digital Food Cultures* (pp. 35–50). London: Routledge.

Simonsen, T.M. (2011). Categorising YouTube. *MedieKultur: Journal of Media and Communication Research*, 27(51): 72–93.

References

Arthurs, J., Drakopoulou, S., & Gandini, A. (2018). Researching YouTube. *Convergence: The International Journal of Research into New Media Technologies*, 24 (1): 3–15.

Barthes, R. (1961). Towards a psychosociology of contemporary food consumption. In C. Counihan & P.Van Esterik (eds.), *Food and Culture* (pp. 28–35). New York: Routledge.

Bärtl, M. (2018). YouTube channels, uploads and views: A statistical analysis of the past 10 years. *Convergence*, 24(1): 16–32.

Berryman, R., & Kavka, M. (2018). Crying on YouTube:Vlogs, self-exposure and the productivity of negative affect. *Convergence*, 24(1): 85–98.

Bishop, S. (2018). Anxiety, panic and self- optimization: Inequalities and the YouTube algorithm. *Convergence*, 24(1): 69–84.

Braun,V., & Carruthers, S. (2020). Working at self and wellness: A critical analysis of vegan vlogs. In D. Lupton & Z. Feldman (eds.), *Digital Food Cultures*. London: Routledge.

Bruns, A. (2008). *Blogs, Wikipedia, Second Life and beyond: From production to produsage.* New York: Peter Lang Publishing.

Bruns, A. (2013). Faster than the speed of print: Reconciling 'big data' social media analysis and academic scholarship. *First Monday*, 18(10). Retrieved November 2019 from: http://ojphi.org/ojs/index.php/fm/article/view/4879/3756

Bryman, A. (2016). *Social Research Methods*. (5th ed). Oxford: Oxford University Press.

Burgess, J., & Green, J. (2009/2018). *YouTube, online video and participatory culture*. Cambridge: Polity Press.

Burroughs, B. (2017). YouTube kids: The app economy and mobile parenting. *Social Media + Society*, April–June, 1–8. Retrieved November 2020 from: https://journals.sagepub.com/doi/full/10.1177/2056305117707189

Cha, M., Kwak, H., Rodriguez, P., Ahn, Y., & Moon, S. (2007). *I Tube, You Tube, everybody tubes*. Paper presented at IMC 07, October 24–26, 2007. Retrieved November 2019 from: https://dl.acm.org/citation.cfm?id=1298309

Chatzopoulou, G., Cheng, S., & Faloutsos, M. (2010). *A first step towards understanding popularity in YouTube*. INFOCOM IEEE Conference on Computer Communications Workshops, San Diego, CA. Retrieved November 2019 from: http://www.cs.unm.edu/~michalis/PAPERS/youtube_CAMERA.pdf

Chen, C. (2016). Forming digital self and parasocial relationships on YouTube. *Journal of Consumer Culture*, 16(1): 232–254.

Chun, E. (2013). Ironic blackness as masculine cool: Asian American language and authenticity on YouTube. *Applied Linguistics*, 34(5): 592–612.

Contois, E.J.H. (2018). The spicy spectacular: Food, gender, and celebrity on Hot Ones. *Feminist Media Studies*, 18(4): 1–4.

Cunningham, S., Craig, D., & Silver, J. (2016). YouTube, multichannel networks and the accelerated evolution of the new screen ecology. *Convergence*, 22(4): 376–391.

Dijck, V.J. (2009). Users like you? Theorizing agency in user-generated content. *Media Culture Society*, 31(1): 41–58.

Dijck, V.J. (2013). *The Culture of Connectivity: A Critical History of Social Media*. Oxford: Oxford University Press.

Douglas, M. (1966). *Purity and Danger*. London. Routledge & Kegan.

Ellison, N.B., & boyd, d. (2013). Sociality through social network sites. In W.H. Dutton (ed.), *The Oxford Handbook of Internet Studies* (pp. 151–172). Oxford: Oxford University Press.

Evans, S.K., Pearce, K.E., Vitak, J., & Treem, J.W. (2016). Explicating affordances: A conceptual framework for understanding affordances in communication research. *Journal of Computer-Mediated Communication*, 22(1): 35–52.

Fischler, C. (1988). Food, self and identity. *Social Science Information*, 27(2): 275–292.

Forchtner, B., & Tominc, A. (2017). Kalashnikov and cooking-spoon: Neo-Nazism, veganism and a lifestyle cooking show on YouTube. *Food, Culture & Society*, 20(3): 415–441.

Fuchs, C. (2012). *Social Media: A Critical Introduction*. London: Sage Publications Ltd.

García-Rapp, F. (2017) Popularity markers on YouTube's attention economy: The case of Bubzbeauty. *Celebrity Studies*, 8(2): 228–245. doi: 10.1080/19392397.2016.1242430

Gibson, M. (2016). YouTube and bereavement vlogging: Emotional exchange between strangers. *Journal of Sociology*, 52(4): 631–645.

Goffman, E. (1959/1990). *The Presentation of Self in Everyday Life*. London: Penguin Books.

Golder, S.A., & Huberman, B.A. (2006). The structure of collaborative tagging systems. *Journal of Information Science*, 32(2): 198–208.

Guo, L., & Harlow, S. (2014). User-generated racism: An analysis of stereotypes of African Americans, Latinos, and Asians in YouTube videos. *Howard Journal of Communications*, 25(3): 281–302. doi: 10.1080/10646175.2014.925413

Halpern, D., & Gibbs, J. (2012). Social media as a catalyst for online deliberation?. *Computers in Human Behavior*, 29: 1159–1168.

Havalais, A. (2013). Home made big data? Challenges and opportunities for participatory social research. *First Monday*, 18(10). Retrieved November 2019 from: http://ojphi.org/ojs/index.php/fm/article/view/4876/3754

Horton, D., & Wohl, R. (1956/2006). Mass communication and para-social interaction: Observations on intimacy at a distance. Reprinted in *Participations – Journal Audience and Reception Studies*, 3(1). Retrieved November 2019 from: http://www.participations.org/volume%203/issue%201/3_01_hortonwohl.htm

Jenkins, H. (2006). *Convergence Culture*. New York: New York University Press.

Kim, K., Paek, H.J., & Lynn, J. (2010). A content analysis of smoking fetish videos on YouTube. *Health Communication*, 25(2): 97–106. doi: 10.1080/10410230903544415

Krippendorff, K. (2013). *Content Analysis : An Introduction to Its Methodology*. Thousand Oaks: Sage Publications.

Lange, P. (2007). Publicly private and privately public: Social networking on YouTube. *Journal of Computer-Mediated Communication*, 13(1): 1–18. Retrieved November 2019 from: http://jcmc.indiana.edu/vol13/issue1/lange.html

Lee, D.Y., & Letho, M.R. (2012). User acceptance of YouTube for procedural learning: An extension of the technology acceptance mode. *Computers & Education*, 61: 193–208.

Lewis, T. (2018). Digital food: From paddock to platform. *Communication Research and Practice*, 4(3): 212–228.

Lewis, T. (2020). *Digital Food: From Paddock to Platform*. London: Bloomsbury Academic.

Lobato, R. (2016). The cultural logic of digital intermediaries: YouTube multichannel networks. *Convergence*, 22(4): 348–360.

Lovelock, M. (2017). 'Is every YouTuber going to make a coming out video eventually?': YouTube celebrity video bloggers and lesbian and gay identity. *Celebrity Studies*, 8(1): 87–103.

Lovink, G. (2008). The art of watching databases. Introduction to the video vortex reader. In G. Lovink & S. Niederer (eds.), *The Video Vortex Reader. Responses to YouTube* (pp. 9–12). Amsterdam: Institute of Network Cultures

Lupton, D. (2020a). Understanding digital food cultures. In D. Lupton & Z. Feldman (eds.), *Digital Food Cultures* (pp. 1–16). London: Routledge.

Lupton, D. (2020b). Carnivalesque food videos: Excess, gender and affect on YouTube. In D. Lupton & Z. Feldman (eds.), *Digital Food Cultures* (pp. 35–50). London: Routledge.

Lupton, D., & Feldman, Z. (eds.) (2020). *Digital Food Cultures*. London: Routledge.

Markham, A.N., & Baym, N.K. (2009). *Internet inquiry: Conversations about Methods*. London: Sage Publications, Inc.
Molyneaux, H., O'Donnell, S., Gibson, K., & Singer, J. (2008). Exploring the gender divide on YouTube. *The American Communication Journal*, 10(2): 1–14.
Pereira, B., Sung, B., & Sean, L. (2019). I like watching other people eat: A cross-cultural analysis of the antecedents of attitudes towards Mukbang. *Australasian Marketing Journal*, 27: 78–90.
Pinto, H. Almeida, J.M., & Goncalves, M.A. (2013), *Using early view patterns to predict the popularity of YouTube videos*. In Proceedings of the Sixth ACM International Conference on Web Search and Data Mining, Rome, Italy, pp. 365–374.
Postigo, H. (2016). The socio-technical architecture of digital labor: Converting play into YouTube money. *New Media & Society*, 18(2): 332–349.
Quennerstedt, M., Flintoff, A., & Webb, L. (2013). Narratives from YouTube: Juxtaposing stories about physical education. *SAGE Open*, 1–10. doi: 10.1177/2158244013507266
Rieder, B., Matamoros-Fernández, A., & Coromina, O. (2018). From ranking algorithms to 'ranking cultures'. *Convergence*: 24 (1): 50–68.
Schönberger, V.M., & Cukier, K. (2013). *Big Data – A Revolution That Will Transform How We Live, Work and Think*. London: John Murray.
Simonsen, T.M. (2011). Categorizing YouTube. *MedieKultur: Journal of Media and Communication Research*, 28 (51): 72–93.
Sinclair, J., & Cardew-Hall, M. (2008). The folksonomy tag cloud: When is it useful? *Journal of Information Science*, 34(1): 15–29.
Smith, D. (2014). Charlie is so 'English'-like: Nationality and the branded celebrity person in the age of YouTube. *Celebrity Studies*, 5(3), 256–274. doi: 10.1080/19392397.2014.903160
Snelson, C. (2013). Vlogging about school on YouTube: An exploratory study. *New Media & Society*, 17(3): 321–339
Stephen, K., & Cumming, G.P. (2010). Searching for pelvic floor muscle exercises on YouTube. *Menopause International*, 18: 110–115.
Strangelove, M. (2010). *Watching YouTube – Extraordinary Videos by Ordinary People*. Toronto: University of Toronto Press.
Thorson, K. Driscoll, K., Ekdale, B., & Edgerly, S. et al. (2013). Youtube, Twitter and the occupy movement. *Information, Communication & Society*, 16(3): 421–451.
Vernallis, C. (2013). *Unruly Media: YouTube, Music Video, and the New Digital Cinema*. Oxford: Oxford University Press
Vonderau, P. (2016). The video bubble: Multichannel networks and the transformation of YouTube. *Convergence*, 22(4): 361–375.
Welbourne, D.J., & Grant, W.J. (2016). Science communication on YouTube: Factors that affect channel and video popularity. *Public Understanding of Science*, 25(6): 706–718.
Whitaker, J., Orman, E., & Yarbrough, C. (2014). Characteristics of "music education" videos posted on YouTube. *Update*, 33(1): 49–56.
Worsley, A., Wang, W., Ismail, S., & Ridley, S. (2014). Consumers' interest in learning about cooking: The influence of age, gender and education. *International Journal of Consumer Studies*, 38: 258–264.
Yoo, J.H., & Kim, J. (2012). Obesity in the New Media: A content analysis of obesity videos on YouTube. *Health Communication*, 27(1): 86–97.

PART 2
Digital ethnography

6

FIELDWORK IN ONLINE FOODSCAPES

How to bring an ethnographic approach to studies of digital food and digital eating

Tanja Schneider and Karin Eli

Introduction to digital food and digital eating

Digital technologies increasingly mediate how producers and consumers seek, share, and interpret food and food-related information and eating practices. In the last few years, scholars have started exploring the growing digitalisation of food and eating and the emerging implications of these novel products, processes, and practices. This literature has theorised the digitalisation of food through the lenses of "digital food cultures" (Lupton, 2018; Lupton & Feldman, 2020), "digital food activism" (Schneider et al., 2018a), and "digital food" (Lewis, 2019). Reviewing this and related literature, we propose to make a distinction between studies of digital food and studies of digital eating. We define studies of digital food as those that focus either on novel food substances or on digitally enabled and data-focused production and distribution processes of foods. We define studies of digital eating as those that focus on novel eating practices enabled and maintained through mobile, sensor-based and digital technologies. Whereas studies of digital food attend to the digitised histories, economies, and trajectories of new food items, studies of digital eating focus on altered subjectivities and the roles that eating and its digitally mediated practices play. Thus, studies of digital food focus primarily on what people eat as a result of digitised and datafied innovation and production processes, while studies of digital eating aim to understand how people eat differently (or not) when using digital technologies, platforms, and devices and the resulting information.

Having distinguished between digital food and digital eating, we suggest that the next step is to conceptualise what "the digital" means in digital food studies. The digitalisation of food and eating is a multidirectional, interactional process. While digitalisation shapes food and eating, multiple actors, platforms, and structures also shape what "the digital" means, is and does. These include, among others, users and developers, digital and analogue devices, food procurement and consumption spaces, and often-unseen policy and research environments. Our chapter therefore reviews methodologies that have been used in recent years to investigate the digitalisation of food, with particular focus on the strengths and limitations of each method in examining "the digital." We conclude with examples from own work on digital food activism. Drawing on these examples, we delineate a proposed ethnographic approach to understanding digital food and digital eating. Such an ethnographic approach, we argue, can problematise reductionist understandings of "the digital," capture the porous boundaries between "digital" and "analogue" food, and reveal the "symbiotic agency" (Neff & Nagy, 2019) of people, food, and platforms.

Challenges of studying digital food and digital eating

As researchers in the social sciences and humanities have shown over the last decades, studying food and eating offers a unique and rich lens on the intersections of self and society. As an emerging phenomenon, digital food and digital eating have been theorised by only a few scholars in the past years. This begs the question: has nothing much changed from predigital to digital times that warrants more research interest both within the field of (critical) food studies and beyond? We suggest that there are currently two main strands of analysis: analysis that emphasises socio-cultural, economic, and political factors shaping food consumption and eating habits in which the digital does not feature prominently; and emerging analysis that emphasises how the social, cultural, economic, and political are inextricably interrelated with processes of digitalisation and datafication (for more detailed overview, see next section). Currently, the overlap between these two strands of analysis is limited, with the exception of the work of selected food and media studies researchers, who have been the first to study how food and eating are represented on social media, blogs, and microblogging platforms (de Solier, 2018; Leer & Povlsen, 2016; Lewis & Phillipov, 2018; Rousseau, 2013) and some early publications on digital food cultures and activism as mentioned in the previous section.

Research interest in this topic is likely to grow and broaden in the future – facilitated by books such as this one – as digitalisation and datafication affect different spaces and practices. For instance, there is an increase in precision or "smart" farming, which utilises sensory devices to collect agricultural big data; the development of blockchain technology to trace food provenance; and the uptake of mobile apps that track users' food consumption or caloric intake. These examples illustrate how food and its production, distribution, and consumption are increasingly translated into digital data. Research on digital food and eating enables in-depth understandings of how people's everyday practices and identities intersect with new and emerging digitally enabled technologies of production, distribution, and consumption, as well as with digital technologies of interaction and communication.

We argue that the general lack of empirical research on the digitalisation of food and eating highlights several challenges attendant to designing studies that explore food and eating in the digital era. From a theoretical perspective, research on digital food and eating requires engagement with relational approaches (e.g.,

Actor–Network Theory, new materialisms, relational geography), which are not yet mainstream in food studies. Conceptually, digital food and digital eating challenges taken-for-granted binaries of the digital and the analogue, consumption and non-consumption. Methodologically, digital food and digital eating, as a constantly evolving phenomenon that spans human actors, food products, and technologies, across different spaces and geographies, necessitates innovative dialogues between quantitative (aggregated) data analysis and qualitative understandings of individuals' lived experiences. Moreover, as digital food and eating involves multiple actors and platforms (which may have little in common with one another, in both ideology and infrastructure), research on digital food and digital eating requires both comprehensive knowledge of the phenomenon as a whole, and the identification of – and granular focus on – platforms and practices that can best elucidate emerging trends. The next section considers how different methodologies have the potential to address these challenges and advocates for an ethnographic approach to digital food and digital eating. We argue that an ethnographic approach is ideally positioned to examine complex socio-technical phenomena situated in multiple systems and settings.

Bringing an ethnographic approach to the study of digital food and digital eating

Across the social sciences, scholars have used diverse methods to study digital food and digital eating. These include network analyses of big data generated on social media platforms, content analyses of user-generated digital media texts, online focus groups and interviews, and participatory action research, among others. We will discuss some of these methods later in the chapter. Our main interest, however, is exploring how to bring an ethnographic approach to digital food studies, and we therefore begin this section with a discussion of ethnography.

What do we mean by an ethnographic approach? There is no single definition of ethnography, underscoring the complexity of this method, as well as the diversity of disciplines that use ethnographic approaches. Broadly speaking, ethnography aims at the detailed inductive investigation of a phenomenon. As such, ethnographic research involves extensive engagement with a selected study site, or with a few selected sites; indeed, Hine (2015b) describes ethnography as a bodily oriented, circumstantially active, and experientially focused methodology. Importantly, ethnography is comprised of multiple data collection methods, including participant observation, semistructured or unstructured interviews, surveys, archival research, and participatory research. This methodological complexity captures the ethnographic view of social phenomena as complex and as situated in multiple systems and contexts (see also Charmaz & Olesen, 1997; Reeves et al., 2008).

Ethnography requires adaptation when studying digital phenomena and practices. However, rather than defining digital phenomena and practices as different from analogue ones and warranting separate methods, Hine advocates for an understanding of the Internet as an embedded, embodied, and everyday phenomenon and suggests that researchers "look both inward and outward in search of ethnographic holism" (Hine, 2015b: 53). By looking outward, ethnographers focus not only on the Internet itself but also on its multiple contexts, embedding the Internet within multiple institutions, devices, and practices of daily life. Hine also encourages ethnographers to look inwards in search of embodied experiences of using the Internet in everyday life. Other researchers have advocated related approaches for doing qualitative digital research in anthropology (Horst & Miller, 2012) and in sociology (Lupton, 2018). Digital anthropologists and sociologists consider digital devices as material culture and users' digital engagements as expressions that reveal shared values and forms of sociality.

How has ethnography contributed to digital food studies? Ethnography has a long tradition in critical food studies, where it has been used by anthropologists, sociologists, and geographers to illuminate socialities of food and eating, relationships between eating and affect, and societal changes in food practices (for an overview see, for instance, Lupton, 1996; Poulain, 2002/2017). These analyses have called attention to networks of eating (Abbots & Lavis, 2013), which include not only individual eaters, but also food producers, distributors, retailers, and regulators, alongside the food products themselves. Through ethnography's focus on lived experiences of eating, food scholars have developed critical accounts of how individual and social eating practices dialogue with macrolevel infrastructures (e.g., Abbots et al., 2015; Cairns & Johnston, 2015; Carolan, 2011). As such, ethnographic work has shown how networks or assemblages of eating extend from the individual, embodied event of consuming food to the wider political–economic structures that shape the landscapes of food growing, manufacturing, and marketing (e.g., Forney et al., 2018; Le Heron et al., 2016).

Despite the important role ethnography has played in critical food studies, ethnographic research in digital food studies is still not widespread. To date, there has been little ethnographic research on the relational co-constitution of food-body-society with and through digital media platforms and devices, with a particularly glaring lack of engagement with the infrastructures of digital media platforms and how they shape emergent eating practices (but see Didžiokaitė et al., 2018b; Ruckenstein, 2015, for studies of dietary or calorie tracking). One exception is the research of sociologist Michael Carolan (2017b), who has recently studied how digital platforms "shape the foodscapes we imagine and enact" (Carolan, 2017b: 818) and also explores the "types of active selves" that interactive assemblages engender based on thick description of two case studies, FarmDrop, a food distribution platform for London-based producers and consumers, and Farm Hack, a US-based group of farmers who promote open code, hack digital locks built into software of farm equipment, and share the information with each other. Carolan defines two types of active selves: active citizens and more-than-active citizens. The latter group, to whom he refers as "activists," aims to disrupt established routines and everyday practices. The digital platform, then, plays an active role in defining particular types of agency around food.

Digital food activism: Digital platforms as actors in networks of eating

Given the paucity of ethnographic work in digital food studies, why do we advocate for an ethnographic approach? Our call for ethnography is inspired by our recently published edited volume, *Digital Food Activism* (Schneider et al., 2018a), which explores digital platforms as influential actors in networks of eating. The volume analyses how emergent forms of food activism – enabled and shaped by digital platforms – remap networks of food politics, production, distribution, and consumption, transforming relationships between consumers and industrial and policy actors. Several chapters in our volume explore these processes through an ethnographic lens, at times zooming in on locales as small as a supermarket or a backyard, to understand larger societal and economic forces at play. These chapters reveal the potential that ethnography holds for digital food studies alongside the challenges involved in developing an ethnographic approach to "the digital." We provide a brief review of ethnographic contributions to our volume before turning to one example from our own research on digital food activism in the next section, to highlight how ethnography has been used to study digital food activism, in comparison to other methods mentioned in the beginning of this section.

In *Digital Food Activism*, the ethnographic chapters focus on illuminating how small groups of consumers, producers, retailers, and hackers engage with digital media to effect change in the food system. Employing a range of ethnographic approaches, these chapters develop nuanced understandings of how digital food activism is enacted in users' everyday practices and encounters. In her chapter on digital food hacktivism, Melissa Caldwell (2018) draws on multisited participant observation as an attendee of food hacking events, based in the United States, Eastern Europe, and Russia. Caldwell explores how a newly emerging cluster of food hackers, citizen-scientists, and entrepreneurs creates disruptive food experiences through digital technologies. She finds that activists' playful and experimental attitudes to food implicate social justice concerns about equity, safety, and access to knowledge, and that digital materialities and spaces allow for the development of participatory food communities that challenge notions of property and expertise. Likewise, Sarah Lyon's (2018) chapter elucidates how people engage with digital technologies to challenge relations of power in the food industry. Drawing on several field methods, including surveys of coffee producers, focus groups, and multiple unstructured interviews conducted across coffee-growing communities, Lyon focuses on how producers in agro-food networks – in this case, growers of fair trade/organic coffee in southern Mexico – use digital technologies to become more competitive in the specialty coffee market. Lyon develops an account of how coffee growers use online record keeping and social media to improve traceability and share their experiences with consumers and funders, arguing that growers adopt/adapt digital media to assert their political agency and promote food sovereignty. In another example, Ryan Foley (2018) draws on a year of intensive participant observation and interviews with an Italian social co-op, to question the boundaries that separate digital food activism from digital food marketing. Having worked with the co-op's marketing team as part of her ethnographic research, Foley explores how the co-op uses Facebook in ways that intertwine products and ideologies, revealing the tensions and paradoxes that imbue the "selling" of conscious consumption.

While Caldwell (2018), Lyon (2018), and Foley (2018) use different methods in their ethnographic studies, their chapters all highlight how an ethnographic approach can provide nuanced understandings of people's complex, sometimes contradictory, engagements with digital media, and how this, in turn, can illuminate digital food activism on a wider scale. However, as other chapters in our volume show, ethnography is not the only methodological pathway towards understanding people's engagements with digital food activism. Eva Giraud's (2018) chapter on anti-capitalist food activism is grounded in participatory action research. Giraud writes that, in contrast to ethnography, participatory action research entails skill-sharing and knowledge/resource co-production, blurring the boundaries that separate activism from research. As an active participant in Veggies, a UK-based grassroots organisation, Giraud finds that producing activist food media implicates power "frictions," particularly now that activists use mainstream commercial social media, like Twitter or Facebook, and that these "frictions" reveal how activists creatively negotiate multiple digital media to reach diverse audiences and consolidate their identity. Along similar lines, Tania Lewis (2018) draws on participatory research with permablitz activists in Melbourne, Australia, to trace the co-production of internet-enhanced food activism across locales, finding that when permablitz groups document their activities online, they provide local models of action that can be adopted by geographically distant communities. Additionally, contributors to our volume have shown how more "traditional" social scientific methods, such as semistructured interviews and focus groups, can be used to understand people's engagements with digital food activism. Drawing on participant diaries and semistructured interviews, Katharina Witterhold (2018) develops an in-depth case study analysis in which she demonstrates how young German women's everyday food practices reveal crucial differences in motivation and political potential between types of consumer activists. Another example is Deborah Lupton and Bethaney Turner's (2018) analysis of an online focus group in which they asked Australian participants to reflect on the possibility of consuming 3D printed foods. Finding that participants discussed 3D printed food as unnatural and potentially abject, Lupon and Turner show how consumers, activists, and technology companies develop definitions of food – what it is, and what it can be – in ways that mobilise specific, and sometimes clashing, values.

As described above, various qualitative methods, including ethnography, participatory action research, focus groups, semistructured interviews, and diary analysis, can help us understand people's everyday engagements with digital food and digital eating. The challenge, however, is how to develop ethnographic approaches that allow us to zoom in on digital platforms themselves and the interactions between users and platforms, and thereby understand the logics and values that underlie the development and operations of digital food infrastructures. Returning to

Digital Food Activism, two chapters that focus on Twitter-based activism stand out in offering possible pathways towards understanding social media platforms. In a chapter that explores how Twitter is used in diabetes-related discourses, Amy McLennan et al. (2018) combine network analysis and iterative thematic coding to understand which Twitter accounts are most influential, and which types of content are tweeted most frequently. Through this mixed methods analysis, the authors explore Twitter as a space where users may share values and coordinate action, while highlighting the lack of transparency about power and influence that inheres in the infrastructure of the platform. In a chapter on Twitter's role in food insecurity activism, Alana Mann (2018) employs the Twitter Capture and Analysis Tool Set (TCAT), to conduct a frame analysis of Twitter coverage of an Australian food security campaign. Mann focuses on hashtags, mentions, and handles as "issue framing devices" and shows how Twitter elements such as hashtags, handles, and mentions can serve as issue-framing devices, arguing that digital media platforms play a role in constituting issues and issue publics. Taken together, these two chapters develop a valuable account of digital food activism that centres on the logics of a social media platform, calling into question the presumed democratisation of knowledge and activism in/through social media.

Digital food activism: The case of Buycott

In our own research on digital food activism, we employed a variety of qualitative methods to examine the co-constitutive agency of people and platforms. To date, our research has focused on three case studies: a wiki platform, a mobile barcode-scanning app, and an online-centric food activism organisation. We have employed different methods to research each case, adapting our approach to the platform and to the organisational context underlying it.

For instance, to research (the pseudonymous) HowToBuyWiki, a non-profit organisation developing an open-source wiki platform dedicated to product transparency, we mainly focused on interviews with the organisation's six active members (Eli et al., 2015). We triangulated these interviews with formal documentation of the organisation's workshops and meetings, provided by the organisation's members, and also spent some time on the platform itself, studying how information was presented both visually and textually. Through this case study, we gained insight into the tensions between consumption and care, democratisation and expertise, that weave into the development of a digital activism platform.

In our case study of foodwatch, a non-profit organisation that uses social media and e-newsletters to promote consumer action around food-related campaigns, we chose to focus on the organisation's main communications tool – weekly e-newsletters (Schneider et al., 2019). Through a thematic analysis of 50 e-newsletters, we explored how foodwatch positions itself in the food policy landscape, identifying interlinked strategies through which foodwatch establishes its own expertise and authority.

Our third case study of Buycott is the focus of this section. Buycott is a barcode-scanning app that promotes political consumption through user-generated campaigns, under the banner of "voting with your wallet."[1] In studying this app, we relied to a considerable extent on textual analysis (Eli et al., 2016). However, our continuous engagement with Buycott enabled us to develop a co-presence that facilitated gaining an understanding of the development of Buycott. The Buycott case study, thus, provides useful insights into the practicalities of studying digital eating, which we describe in more detail in the next paragraphs.

Initially, our research of Buycott involved participatory engagement with the app. In 2013, following media coverage of Buycott's campaigns for the labelling of genetically modified (GMO) foods and against Koch Industries, we decided to study how consumers engage with Buycott in everyday decision-making about food. We began by joining Buycott campaigns and using the app to scan the products we bought or already had in our homes or offices in Switzerland and the UK, respectively. These auto-ethnographic explorations proved crucial for gaining first-hand experience of learning how to do app-mediated political consumerism. It also led to many questions (e.g., how exactly does the app work and based on which data? Who checks that the user-generated information is correct?), interesting revelations (e.g., a small Swiss mineral water producer being owned by a US headquartered beverage corporation), and frustrations (e.g., certain products not being recognised and the request to manually add them to the database). It also occasionally led us to reflect about our consumption practices – as one of us described in a short blog post (Schneider, 2013) – thus leading us to engage in specific boycotts and buycotts, as the script of the app intended. On other occasions, though, the app provided us with ambiguous, unreliable, and conflicting information, which was of little practical use.

Our study continued, and in parallel to our own participatory insights about Buycott, we found that, on social media and app review sites, consumers were employing discourses that diverged from Buycott's own (stated) project. In 2014, our focus shifted to exploring how consumers understand the ethical, political, and informational ramifications of Buycott. As we described in our article on Buycott (Eli et al., 2016), although Buycott's project was envisioned – and discursively constructed – as participatory, this participatory script did not apply to many users, who engaged with the app as recipients, rather than contributors, of information, leading to conceptual and practical gaps. Later in 2014, a new aspect of Buycott emerged, and our case study shifted again. In the wake of the summer war between Israel and Gaza, a Buycott campaign titled "Long live Palestine boycott Israel" came to the foreground of both media and user attention.

Through following the unfolding of this campaign, we recognised that the case of Buycott was no longer about the mismatch of users and scripts, but about "dynamic co-constitution, involving the triad of the news media, citizen-consumers, and the ICT platform" (Eli et al., 2016: 66). We therefore decided to focus our case study on Buycott's two defining campaigns – "Long live Palestine" and "Demand GMO Labelling" – and collect data focused on the triad we identified: all news media texts (published online, April 2013–August 2014), user-generated posts (Buycott Facebook page, iTunes user reviews), and texts generated by Buycott's official team (the Buycott's website, Facebook page, Twitter accounts). We conducted a thematic discourse analysis of these texts, with particular interest in understanding tensions between media and user framings of Buycott and the identity that Buycott's developers constructed for the app. Through this approach, we developed an analysis of how multiple discourses co-construct and also delimit "moments of possibility" (Barnes, 2017) for consumer action. In our paper based on this analysis, we explored how, in dialogue with Buycott's focus on individual consumer action conducted through barcoded items in retail spaces, users' discourses materialised political issues within products. For example, when referring to the "Demand GMO Labelling" campaign, many users wrote that Buycott informed them about products that contained GMO, thus translating a campaign for GMO labelling as a campaign that exposed "toxic food" (Eli et al., 2016: 70). Likewise, when writing about the "Long Live Palestine" campaign, many users imagined that boycotted products were directly linked to Israeli land or were used to fund weapons. In both cases, we found that users substantiated campaigns on Buycott through tangible consumer products, thereby transforming the app itself into an ethical commodity to be consumed.

As we wrote up our findings about all three case studies and compared the different cases, new questions related to Buycott emerged – questions we were able to answer neither based on further media or website analysis nor based on our participatory engagement with the app. The questions centred on Buycott sources of funding, its plans for the future and how Buycott and its team members dealt with information that consumers entered about products and the campaigns consumers launched. For instance, how did the Buycott team verify, whether the information was correct? Were all campaigns approved and did the Buycott team check whether the "to-be-boycotted" products or companies were correctly linked to specific campaigns? This led us to approach the start-up and request an interview with the founder, Ivan Pardo, in early 2017, which he agreed to give via Skype. Asking him about the consumers' and producers' reactions after the launch of the app, he shared with us that consumers' reactions were always positive, whereas "the response from the brands was more or less silence" and that companies rarely engaged with them directly. Pardo stated, "I know that they know that we exist but they mostly don't acknowledge us." However, there was one exception: brands occasionally contacted Buycott by email if some of the user-generated information was not correct, asking for a correction. Overall, the interview provided answers to our questions (cf. Eli et al., 2018) and complemented our previous insights, however, the question we got asked most during conference presentations – does boycotting/buycotting have an impact? – is a question we still could not provide a conclusive answer to; perhaps in part due to the brands' silence as a response to the consumers' actions.

Two years later, when one of us was preparing an academic talk that drew on our research on Buycott, we checked back to find that the Buycott website and mobile app had been updated. The main change we noticed was a new tab on the website entitled "Barcode API." A click on the tab led to the discovery that Buycott advertises itself as providing access to "The world's largest UPC database," stating that "our comprehensive product API provides data for over 150 million products from every corner of the globe."[2] A UPC (Universal Product Code), otherwise known as a barcode, offers a unique electronic identifier for products, allowing for scanning and tracking items in the retail sector. Buycott now offers paid plans to access this database: Plan Basic for $49 a month, Plan Developer for $99 a month and Plan Startup for $499 a month. So, the crowd-sourced data that activist-consumers enter to create campaigns and connect brands/products to these products are now of monetary value to businesses, including Buycott itself. When we interviewed Ivan Pardo in 2017, he did not raise the possibility that Buycott might sell access to its vast database. Now it seems that Buycott's crowd-sourced data serve as the basis for a business model, likely to increase Bucyott's financial valuation. The consumer-activists entering the data contribute to this valuation; however, their activities, enabled by digital and mobile technologies, are uncompensated. This blurs the boundaries between consumption and production, and one may argue that Buycott users provide free "digital labour," typical of the digital economy (cf. Scholz, 2013).[3]

The development we chart above is signified in the succession of company logos that Buycott has been using since it was founded in 2013. When we first engaged with the app, the logo we encountered showed a clenched white fist holding three green dollar bills ($20 visible on the first) against a black background. Over time, this fist changed and became more abstract, holding a light green slip of paper that could connote not only a dollar bill but also a paper note or ballot paper. The background also changed from black to turquoise. In the logo's latest update, the fist and bills have disappeared. The current logo depicts a white lower-case b against a red background, placing the Buycott brand – rather than the consumer "empowerment" it claims to facilitate – centrestage.

Based on our research on digital food activism, and the Buycott case in particular, we propose that ethnographic research drawing on different methodological elements, such as autoethnography, participant observation, semistructured interviews, and media analysis, enables co-presence with digital platforms and their interfaces, devices, users, and objects over a longer periods of time. This co-presence provides insights into the platformisation of eating, the enactment of data assemblages and shifting accountability relations (cf. Schneider et al., 2018b). When researching the digital economy and digitally enabled practices, long-term fieldwork, which is typical of ethnographic research, may be re-interpreted to mean longer-term ethnographically inclined engagement with emerging, evolving and elusive digital technologies.

Lessons learned and outlook

There could be a temptation to approach digital food and digital eating with methodologies that explicitly give priority to exploring the "digital" nature of the chosen phenomena and practices. For example, the research problem of understanding digital food activism could be tackled by employing digital ethnography. But is it that simple? Our research and that of others in this area problematises such a narrow understanding of the digital in digital food studies that exclusively attends to the digital by digital means. Hine (2015b) emphasises how the online/offline divide is in practice not always separate but co-present, and hence argues that researchers should ideally integrate the analysis of online engagements with "real world" interviews and observations. In particular, Hine urges a cautious interpretative stance, which acknowledges that, in relying on online discourses, we do not have access to the particular spatio-temporal, material, or embodied contexts that framed each user's, journalist's, or the app developer's discursive expressions.

A second lesson is that researchers need to reflect about and address the potential issue of how to manage "ethnographic distance" before, during, and/or after fieldwork. For example, researchers may regularly use food delivery apps to order food, self-tracking devices to monitor their diet, or post food photos on Instagram in their everyday life. These regular users of digital media face the challenge of rendering the familiar strange, which is essential in ethnographic research. Likewise, researchers who rarely use mobile apps, digital devices, or social media platforms in their food-related practices face an equally challenging ethnographic task: to render the strange familiar. Ultimately, we would like to emphasise the importance of researchers stepping out of their comfort zone to defamiliarise themselves and gain a new understanding of a phenomenon.

A third lesson learned is the need for more studies of digital food and eating. The expected increase in digitalisation and datafication affecting food requires social science expertise on these processes and their effects on food and eating. An ethnographic approach can contribute an in-depth understanding of emerging practices and the changing values attendant to them. Moreover, multisited ethnographies of the same phenomenon add to a broader understanding of how the same processes of digitalisation and datafication vary but are also related dependent on context (including legal, political, and cultural) and how they are co-constituted. In general, this independent academic expertise is pivotal, we would argue, in the face of a growing corporate digital food economy. Given the prevailing data ownership and use strategies employed by corporate actors, it is likely that digital data will not be shared with consumers or with academic researchers. Data providing insights into what and how people eat could thus be in the hands of a few food producers, restaurants, and retailers – or most likely Amazon – (and those data analysis companies they sell their data to). Social science researchers can therefore play a crucial role in developing alternative (open access) data sources, and interpretations thereof.

We suggest that the next step for digital food studies is to elucidate how the multiple actors, platforms, and structures involved in digital food and eating shape what "the digital" means and does. These include, among others, users and developers, digital and analogue devices, food procurement and consumption spaces, and often-unseen policy and research environments. Such an ethnographic approach, we would like to emphasise once more, can problematise reductionist understandings of "the digital," capture the porous boundaries between "digital" and "analogue" food, and reveal the complex "symbiotic agency" (Neff & Nagy, 2019) of people, food, and platforms. An ethnographic lens on digital food and digital eating would thus bring this emerging field into productive dialogue with broader socio-technical and new materialist theories of food and eating.

Dos and Don'ts

Do step out of your comfort zone and seek to defamiliarise yourself, to make sure you have a fresh perspective on the phenomenon you're studying.

Don't assume that data provided by companies are value free.

Do take screenshots and make sure that websites are archived.

Do craft a statement that establishes the value of your research – participants should feel that they have a stake in research that matters.

Don't assume people will want to talk to you; spend time researching and connecting, preparing your field.

Do reflect on ethics – not all data collected online are subject to the same ethical considerations.

Now you try

A cornerstone of ethnography is participant observation. This is also one of the most difficult methodologies to teach and to practice. One way of gaining some experience in participant observation within digital food cultures is to follow a food hashtag within a social media platform, such as Instagram or Twitter. Many food hashtags are associated with particular digital food communities, so choosing a hashtag to follow can give you a lens onto how a digital food community represents itself and its food practices, communicates shared values both visually and textually, and reinforces particular norms around food and eating. For this activity, we encourage you to choose a hashtag associated with a food community whose practices or ideologies are mostly unfamiliar to you. Then, choose a social media platform and search for the hashtag. For an hour or two, follow the hashtag and take notes: what words or phrases appear frequently in posts that include this hashtag? What types of images are associated with the hashtag? Does the hashtag appear alongside other hashtags? What tone do the posts have? Are the posts overwhelmingly positive, or do you encounter some debates, and if so, which points of view seem to be at odds? Once your one or two hours of observation are completed, revisit your notes and try to write a short narrative, summarising what you've learned about the hashtag, and what new questions have emerged through your observations.

Questions for reflection

What is the difference between digital food and digital eating and what is the focus of each subfield of digital food studies?

What is the difference between ethnography, digital ethnography, and an ethnographic approach to digital food and digital eating?

What did you learn about digital food activism, as researched through an ethnographic approach, that surprised you? What did you find particularly insightful?

Ethical considerations

Ethnographic research, both analogue and digital, carries several ethical considerations, of which we will delineate a few key areas. Given the complexity of ethnographic research, which relies on multiple sources of data – including participant observation, interviews, archived images or texts, and surveys, among others – it is important to consider the ethical concerns specific to each of these methods. Following a popular food hashtag on a social media platform would not usually require that you seek informed consent, but if you choose to follow a hashtag specific to a small or potentially vulnerable community of users – for example, an online community of people with eating disorders – you will likely need to tell the users about your research and seek community consent to proceed. In another example, although interacting on social media platforms via replies, mentions or direct messaging is often easy and intuitive, if you intend to use these interactions in your research, it is important to remind participants about your study and ask if you can use these interactions. Moreover, an interaction on social media may become the basis for a more formal interview. In this case, even if your interview is conducted online, the participant will still need to provide verbal or written informed consent for the interview to take place. Other ethical queries you will likely encounter during your research will relate to interactions with textual or visual data published by users. Even where these data are published publicly (and not as part of restricted accounts or closed groups), you may need to consider how to use the data on a case-by-case basis, especially if the confidentiality of participants or groups might be compromised by the inclusion of these data in research and, particularly, publication. For additional discussion of the ethics of using social media and blog posts in research, please refer to Hookway (2008) and Lavis (2017).

Further readings

Carolan, M. (2017). Agro-digital governance and life itself: Food politics at the intersection of code and affect. *Sociologia Ruralis*, 57(S1): 816–835.

Didžiokaitė, G., Saukko, P., & Greiffenhagen, C. (2018a). The mundane experience of everyday calorie trackers: Beyond the metaphor of quantified self. *New Media and Society*, 20(4): 1470–1487.

Hine, C. (2015a). *Ethnography for the Internet: Embedded, Embodied and Everyday*. London: Bloomsbury.

Schneider, T., Eli, K., Dolan, C., & Ulijaszek, S. (eds.) (2018). *Digital Food Activism*. London: Routledge.

Notes

1 For more information about Buycott see their website: https://buycott.com.

2 https://www.buycott.com/api (accessed 20 January 2020).

3 Other forms of data collection might occur when using the Buycott website (e.g., tracking) or when using the app, as users need to log into the app.

References

Abbots, E.-J. & Lavis, A. (eds.) (2013). *Why We Eat, How We Eat: Contemporary Encounters Between Foods and Bodies*. Farnham: Ashgate.

Abbots, E.-J., Lavis, A., & Attala, L. (eds.) (2015). *Careful Eating: Bodies, Food and Care*. Farnham: Ashgate.

Barnes, C. (2017). Mediating good food and moments of possibility with Jamie Oliver: Problematising celebrity chefs as talking labels. *Geoforum*, 84: 169–178.

Cairns, K., & Johnston, J. (2015). Choosing health: Embodied neoliberalism, postfeminism, and the 'do-diet. *Theory and Society*, 44: 153–175.

Caldwell, M. L. (2018). Hacking the food system: Re-making technologies of food justice. In: Schneider, T., Eli, K., Dolan, C., & Ulijaszek, S. (eds.), *Digital Food Activism* (pp. 25–42). London: Routledge.

Carolan, M. (2011). *Embodied Food Politics*. Farnham: Ashgate.

Carolan, M. (2017b). Agro-digital governance and life itself: Food politics at the intersection of code and affect. *Sociologia Ruralis*, 57(S1): 816–835.

Charmaz, K., & Olesen, V. (1997). Ethnographic research in medical sociology: Its foci and distinctive contributions. *Sociological Methods and Research*, 25(4): 452–494.

de Solier, I.. (2018). Tasting the digital: New food media. In K. LeBescov & P. Naccarato (eds.), *The Handbook of Food and Popular Culture* (pp. 54–65). London: Bloomsbury.

Didžiokaitė, G., Saukko, P., & Greiffenhagen, C. (2018b). The mundane experience of everyday calorie trackers: Beyond the metaphor of quantified self. *New Media and Society*, 20(4): 1470–1487.

Eli, K., Dolan, C., Schneider, T., & Ulijaszek, S. (2016). Mobile activism, material imaginings, and the ethics of the edible: Framing political engagement through the Buycott app. *Geoforum*, 74: 63–73.

Eli, K., McLennan, A. K., & Schneider, T. (2015). Configuring relations of care in an online consumer protection organization. In E.-J. Abbots, A. Lavis, & L. Attala (eds.), *Careful Eating: Bodies, Food and Care* (pp. 173–194). London: Routledge.

Eli, K., Schneider, T., Dolan, C., & Ulijaszek, S. (2018). Digital food activism: Values, expertise and modes of action. In T. Schneider, K. Eli, C. Dolan, & S. Ulijaszek (eds.), *Digital Food Activism* (pp. 203–219). London: Routledge.

Foley, R. A. (2018). Marketing critical consumption: Cultivating conscious consumers or nurturing an alternative food network on Facebook?. In T. Schneider, K. Eli, C. Dolan, & S. Ulijaszek (eds.), *Digital Food Activism* (pp. 110–129). London: Routledge.

Forney, J., Rosin, C., & Campbell, H. (eds.) (2018). *Agri-Environmental Governance as an Assemblage Multiplicity, Power, and Transformation*. London: Routledge.

Giraud, E. (2018). Displacement, 'failure' and friction: Tactical interventions in the communication ecologies of anti-capitalist food activism. In T. Schneider, K. Eli, C. Dolan, & S. Ulijaszek (eds.), *Digital Food Activism* (pp. 130–150). London: Routledge.

Hine, C. (2015b). *Ethnography for the Internet: Embedded, Embodied and Everyday*. London: Bloomsbury.

Hookway, N. (2008). 'Entering the blogosphere': Some strategies for using blogs in social research. *Qualitative Research*, 8(1): 91–113.

Horst, H., & Miller, D. (eds.) (2012). *Digital Anthropology*. London: Berg.

Lavis, A. (2017). Food porn, pro-anorexia and the viscerality of virtual affect: Exploring eating in cyberspace. *Geoforum*, 84: 198–205.

Le Heron, R., Campbell, H., Lewis, N., & Carolan, M. (eds.) (2016). *Biological Economies: Experimentation and the Politics of Agri-Food Frontiers*. London: Earthscan.

Leer, J., & Povlsen, K. K. (eds.) (2016). *Food and Media: Practices, Distinctions and Heterotopias*. London: Routledge.

Lewis, T. (2018). Food politics in a digital era. In: Schneider, Tanja, Eli, Karin, Dolan, Catherine and Ulijaszek, Stanley (eds.), *Digital Food Activism* (pp. 185–202). London: Routledge.
Lewis, T. (2019). *Digital Food: From Paddock to Platform*. London: Bloomsbury.
Lewis, T., & Phillipov, M. (2018). Food/media: Eating, cooking, and provisioning in a digital world. *Communication Research and Practice*, 4(3): 207–211.
Lupton, D., (1996). *Food, the Body and the Self*. London: Sage.
Lupton, D., (2018). Cooking, eating, uploading: Digital food cultures. In K. LeBesco & P. Naccarato, (eds.), *The Handbook of Food and Popular Culture* (pp. 67–75). London: Bloomsbury.
Lupton, D., & Feldman, Z. (eds.) (2020). *Digital Food Cultures*. London: Routledge.
Lupton, D., & B. Turner (2018) 'Both fascinating and disturbing': Consumer responses to 3D food printing and implications for food activism. In T. Schneider, K. Eli, C. Dolan, & S. Ulijaszek (eds.), *Digital Food Activism* (pp. 151–167). London: Routledge.
Lyon, S. (2018). Digital connections: Coffee, agency, and unequal platforms. In T. Schneider, K. Eli, C. Dolan, & S. Ulijaszek (eds.), *Digital Food Activism* (pp. 70–88). London: Routledge.
Mann, A. (2018). Hashtag activism and the right to food in Australia. In T. Schneider, K. Eli, C. Dolan, & S. Ulijaszek (eds.), *Digital Food Activism* (pp. 168–184). London: Routledge.
McLennan, A. K., Ulijaszek, S., & Beguerisse-Díaz, M. (2018). Diabetes on Twitter: Influence, activism and what we can learn from all the food jokes. In T. Schneider, K. Eli, C. Dolan, & S. Ulijaszek (eds.), *Digital Food Activism* (pp. 43–69). London: Routledge.
Neff, G. & Nagy, P. (2019). Agency in the digital age: Using symbiotic agency to explain human–technology interaction. In Z. Papacharissi (ed.), *The Networked Self: Human Augmentics, Artificial Intelligence, Sentience* (pp. 97–107). London: Routledge.
Poulain, J.-P. (2002/2017). *The Sociology of Food: Eating and the Place of Food in Society*. London: Bloomsbury.
Reeves, S., Kuper, A., & Hodges, B. D. (2008). Qualitative research methodologies: Ethnography. *British Medical Journal*, 337: a1020.
Rousseau, S. (2013). *Food and Social Media: You Are What You Tweet*. Plymouth: AltaMira Press.
Ruckenstein, M. (2015). Uncovering everyday rhythms and patterns: Food tracking and new forms of visibility and temporality in health care. In L. Botin, P. Bertelsen, & C. Nøhr (eds.), *Techno-Anthropology in Health Informatics: Methodologies for Improving Human-Technology Relations* (pp. 28–40). Amsterdam: IOS Press.
Schneider, T. (2013, August 20). *Learning how to buycott? Political consumerism and new media* [web log post]. Retrieved from https://oxfordfoodgovernancegroup.wordpress.com/2013/08/20/learning-how-to-buycott-political-consumerism-and-new-media/
Schneider, T., Eli, K., Dolan, C., & Ulijaszek, S. (eds.) (2018a). *Digital Food Activism*. London: Routledge.
Schneider, T., Eli, K., Dolan, C., & Ulijaszek, S. (eds.) (2018b). Digital food activism: Food transparency one byte/bite at a time. In T. Schneider, K. Eli, C. Dolan, & S. Ulijaszek (eds.), *Digital Food Activism* (pp. 1–24). London: Routledge.
Schneider, T., Eli, K., McLennan, A., Dolan, C., Lezaun, J., & Ulijaszek, S. (2019). Governance by campaign: The co-constitution of food issues, publics and expertise through new information and communication technologies. *Information, Communication & Society*, 22(2): 172–192.
Scholz, T. (ed.) (2013). *Digital Labor: The Internet as Playground and Factory*. New York: Routledge.
Witterhold, K. (2018). Political consumers as digital food activists?: The role of food in the digitalization of political consumption. In T. Schneider, K. Eli, C. Dolan, & S. Ulijaszek (eds.), *Digital Food Activism* (pp. 89–109). London: Routledge.

7

BEYOND THE HASHTAG

Social media ethnography in food activism

Alana Mann

Social media are sites of rich cultural processes with the potential to promote collective action. The practices of food activists are increasingly shaped through the affordances of digital media platforms where they engage in "reflexive forms of activism" (Caroll & Hackett, 2006: 96; Maireder & Asserhofer, 2014) though digital campaigns that "link speech [with] physical action" (Chouliaraki, 2013: 277). Ordinary people not only forge personalised patterns of news consumption through social media platforms (Couldry, 2000; Bruns, 2008; Kwak et al., 2011) but, by sharing and commenting on news stories, "frame and potentially change news agendas" (Jensen, et al., 2016: 8). Further, storytelling through processes of "trans-mediation" – the translation of ideas, concepts, and creative works into different mediums – is enhanced by the affordances or "action possibilities and opportunities" (Faraj & Azad, 2012) of new multimedia technologies which make the creation of rich, sharable content available to everyday users.

The affordances perspective, as a relational approach to understanding how we interact with technology, is "multifaceted," both enabling and constraining user behaviours in a specific context. Affordances emerge in "mutuality between those using technologies, the material features of technologies, and the situated nature of use" (Evans et al., 2017: 36). Hashtags are affordances of the Twitter platform that "organise instances of momentary connectedness into networks" through, firstly, their generation and, secondly, through their capacity to generate actions including expression, discussion, and engagement (Rathnayake & Suthers, 2018). The notion of momentary connectedness is useful in that it describes how issue-response hashtags are "momentary acts" that connect users who may, or may not, intend to create or interact with a defined community or public.

For the researcher social media platforms and affordances such as hashtags offer "not necessarily new techniques or fora but rather new ways of asking questions, generating insights, and forging communities" (Caldwell, 2018: 39). Online digital tools and methods also offer "critical and creative ways to analyse and visualise 'issue dynamics'" in relation to vital concerns such as climate change (Marres & Gerlitz, 2016). These emergent lines of enquiry call for multi-site, multi-media, and user-centred methodological approaches that attach digital traces to their contexts. In this chapter, I share my first experience of engaging in social media analysis. I begin by explaining the function of hashtags as issue-framing devices. I go on to describe an empirical study of the Australian #righttofood advocacy campaign, conducted on Twitter in 2014. The study reveals interactions that might lead to the emergence of "unanticipated coalitions" (Rodak, 2019) and examples of convergent frames that problematise the objectives of the original campaign. I conclude the chapter with a summary of learnings and ethical considerations that emerged from the study.

#hashtagactivism

Hashtags allow users to cluster, share, and modify messages; as such they facilitate the networking of actors engaged in digital activism or campaigning. Organising content along themes and keywords, hashtags are "a means of coordinating a distributed discussion between more or less large groups of users, who do not need to be connected through existing 'follower' networks" (Bruns & Burgess, 2011: 1). They are "explicit attempts to address an imagined community of users" and have the potential to "act as a bridge" between the hashtag community and the user's network (Bruns & Burgess, 2011: 4). Unregulated and unclassified, hashtags are popular for their "cultural generativity" (Burgess, 2011). Ad hoc or intimate publics can and do emerge around hashtags which, by marking out the intended meaning of an utterance, signpost ["what a statement is really about" (Bonilla & Rosa, 2015)]. As an example of "scalable sociality" (Miller et al., 2016), the hashtag can be addressed to a specific in-group or remain obscure to strangers. Multilingual tags are useful for sociolinguistic analysis as they reflect a language situation, while hashtags accompanying images present opportunities for studying practices including vernacular visual culture on platforms such as Instagram, where the Explore Location facilitates researcher access to data place-based storytelling, from which "traces of real-life contexts emerge as part of a whole picture" (Utekhin, 2017: 192). Tagging places rhetorically centres a physical site – village, town, or city – as protagonist; it may demonstrate pride or condemnation but either way "reminds us of how a visitor leaves a trace in the form of an inscription on a wall stating her home place" (ibid.: 196). As such, hashtags offer statistical overviews and access to rich content that can provide a foundation for ethnographic work.

Hashtag activism, coined by New York Times journalist David Carr (2016), happens when large numbers of postings appear on social media under a common hashtagged word, phrase, or sentence with a social or political claim. Prominent examples include #MeToo (Wood, 2018), #Ferguson (Bonilla & Rosa, 2015), #BringBackOurGirls (Ofori-Parku & Moscato, 2018), and #BlackLivesMatter (Yang, 2016). Political slogans are part and parcel of social movement culture, therefore hashtag activism is better understood as "hashtag-related activism" (Ofori-Parku

& Moscato, 2018: 2482; my italics). As native technological features that unite conversations across multiple platforms, hashtags are intertextual and interdiscursive. By indexing or semantically aggregating stories that are otherwise "dispersed digital discourses" (Airoldi, 2018), they enable researchers to conduct meta-fieldwork across multiple un-sited fields. These data have its own meta-based logic as disconnected self-narrations on a particular topic or issue create a user-generated trail that traces the outlines of a metaphor or narrative.

Narrative conventions vary from platform to platform; "everyday" hashtags rarely evolve from instances of momentary connectedness into collective events. Alice Daer et al. (2014) identify the following "rhetorical functions" of hashtags: emphasising, identifying, critiquing, iterating, and rallying. Rauschnabel et al. (2019) uncover amusing, organising, designing, conforming, trendgaging (trending + engaging), bonding, inspiring, reaching, summarizing, and endorsing. These functions enable users to share their multiple perspectives on political issues (Yardi & boyd, 2010) and serve as a way to frame and structure debates (Bruns, 2012). Replete with action verbs that petition, demand, appeal, protest, refuse, object, and offer ironic interpretations, hashtag activism can mobilise communities around a social issue or cause and enable campaigners to circumvent conventional, and often unhospitable, media venues. They may avoid real or perceived bias and sidestep frame "traps" (Goffman, 1974). Conversely, the brevity of hashtag leaves it short on nuance and high on ambiguity which, given the diverse motivations of tweeters, can lead to misframings and misinterpretations (Bonilla & Rosa, 2015).

Hashtags as framing devices

Within topical conversations marked by hashtags and trending topics, influencers and gatekeepers can be identified, enabling the researcher to map the network topology through social network analysis, and to explore hashtags as framing devices rather than merely "lightweight, semantic annotations that publics assign to tweets in their efforts to self-tag generated content" (Meraz & Papacharissi, 2016: 104). Frame analysis explores how patterns of discourse are established and how they mobilise publics. The verification of frames is made through the close study of campaign discourse, the "stuff of framing" (Steinberg, 1998: 862) that enables activists to share a common interpretation of reality. Understanding individual frames, or "schemata of interpretation" (Goffman, 1974), becomes the essence or raison d'etre of a campaign as a process of issue construction. Networked framing is described by Zizi Papacharissi as a "process through which particular problem definitions, causal interpretations, moral evaluations, and/or treatment recommendations attain prominence through crowd-sourcing practices" (2015: 75). Twitter user practices, including the ability to deploy hashtags and retweet, can facilitate frame setting and frame building (Meraz & Papacharissi, 2013). Twitter users engage in networked framing with advocacy organisations as they revise, rearticulate, and redisperse frames. Symbiotic, conversational interactions on topics of mutual interest between users on social media platforms facilitate the performance of identity (Goffman, 1959) and participation in processes of frame negotiation. This "connective action" embodies self-motivated participation through the sharing of personal content that may be co-produced and co-distributed among peers, non-government organisations (NGOs) and other groups with congruent ideas, plans, resources, and networks (Bennett & Segerberg, 2012; Benkler, 2006).

Critiques of hashtag campaigns as "ephemeral, lacking context, and emblematic of public inaction" (Ofori-Parku & Moscato, 2018: 2481) abound. Accordingly, it is important that the capacity of affordances within technology to affect structural change are interrogated. In the case of #BringBackOurGirls, the campaign to free 276 young women kidnapped by the Boko Haram group in Nigeria in 2014, hashtags serve as "interaction enablers" and represent "contra-flow from the global periphery" (Ofori-Parku & Moscato, 2018: 2494) but "institutional norms, local politics and contextual realities can constrain social media activists' preferred framing of important issues" (ibid., 2018: 2480). Yet, undisputedly, hashtags facilitate the diffusion of information, validate comments, express agreement, and offer support for positions on important issues. Hashtag networks, as examples of structures of connectivity, may be momentary yet, as interpreted and made visible by Twitter's infrastructure, they can mobilize and focus collective action (Rathnayake & Suthers, 2018). Accordingly they are an important strategic resource for NGOs and other civil society groups, as the following case study of hashtag use in a Twitter campaign illustrates.

Hashtag analysis on Twitter

Twitter has evolved from an "awareness system" (Kwak et al., 2011) to "a site for resistance" (Williams, 2015). The study of "Twitter II" as an object seeks to "de-banalise" (Weller, et al., 2014) the platform by highlighting its value as a way to study cultural conditions and trace the evolution of an issue through "professional communities of practice" (Turoff & Hiltz, 2009). This epistemological shift reflects the evolution of the platform from inconsequential, friendly tweets to event-following, advocacy, and mobilisation. Analysis of Twitter data can reveal networking topics, actors, and media objects (Maireder & Asserhofer, 2014) and also lends itself well to textual analysis, including co-word analysis (Marres & Weltevrede, 2013). Identification of the URLs most shared in a period of analysis reveals networks of associations between media objects as mediators on issues, the relationships between discursive frames operating, and "subject drift" away from the original topic. A content-oriented perspective provides insight into shared meanings, ideologies, understandings, discourses, and norms. Issue networks can be seen to merge via co-located hashtags – those occurring side by side in the same tweet – which provide subject matter classification (networking topics). Retweets identify significant posts and experts in the field (networking actors), while replies and followers facilitate network analysis. Unidirectional (one way) and bidirectional (two way) connections between individual users, media organisations, and civil society groups can identify intentions, motivations, common interests, cliques, sub-conversations, and spawning coalitions. The networks of association between users, hashtags, and media objects such as URLs illustrate a wide range of discursive frames at play and demonstrate the synergies between local campaigns on a transnational scale. The scalability of these networks, and their capacity to supplement traditional activist communication and linking practices, enables the members of advocacy organisations to connect with other activists in diverse geographic locations and extend the reach of a campaign or movement, as in the following case.

Case study: The #righttofood campaign

In 2016, members of the Australian Right to Food (RtF) Coalition, a "national food security workforce" (Right to Food Coalition, 2016), attempted to drive collective action on food insecurity via the micro-blogging social networking platform Twitter. As a participant-observer, I conducted a frame-critical analysis of the campaign using the Twitter Capture and Analysis Tool Set (TCAT). The study provided insight into how Twitter elements such as hashtags, handles, and mentions can serve as issue-framing devices and mediators within food advocacy networks when taken up by multiple users and appearing in a "mutually visible space" (Rathnayake & Suthers, 2018). It demonstrated the strategic value of the tool for facilitating a cross-flow of information between ideologically aligned advocacy organisations, both domestically and internationally. These affordances contrasted with its limitations in facilitating political conversation and participation.

Through the Twitter campaign, the RtF Coalition aimed to raise public awareness of domestic food insecurity and the gaps in social provisioning in Australia. In doing so, it promoted the long-term aspiration to "shift the debate from food poverty and charity to one demanding food as a fundamental human right" (Lindberg et al., 2016: 5; Food and Agriculture Organisation of the United Nations (FAO), 2004). The Coalition staged the focused Twitter campaign over three days (June 21–23, 2016) in the lead up to the Australian Federal Election. The campaign had two main aims: to elevate the issue of food insecurity among Australians by targeting elected members across a range of portfolios, both Ministers and their Shadows, and to increase the membership of the RtF Coalition. The strategy was to frame the conversation on food insecurity through prefabricated "organisationally promoted frames" in messages that were "personalised, yet at the same time generic in nature" (Hopke, 2015: 9). Core activists generated the messages and hence the framings, engaging in a form of "centralised strategic management" (Bennett & Toft, 2009: 252). These messages were amplified through organisational and individual supporter accounts including the @WePublicHealth Twitter account managed by Croakey, an independent journalism platform on public health issues which I curated as a member of the RtF Coalition for the period of the campaign. Those of us with active Twitter accounts were contacted directly via email and encouraged to collectively post questions such as Why do we talk about people not having affordable housing but not about people having affordable healthy food? Join @right_to_food (Wednesday June 22, 2016).

Scraping the platform using TCAT, a freely available platform of publicly available data comprised of millions of tweets, enabled me to build an issue inventory based on Twitter posts including the hashtag #righttofood. TCAT provides "robust and reproducible data capture and analysis," easy downloading and methodological transparency (Borra & Rieder, 2014: 263). I performed basic data processing to set up TCAT for extracting URLs, hashtag #righttofood and @mentioned accounts @right_to_food belonging to the Right to Food Coalition, and @WePublicHealth. I then used TCAT to sort tweets posted over the three-day period of the campaign into the following categories: straightforward tweets, replies/mentions, and retweets. User statistics by individual, hashtag frequency, user mention frequency, URL, media, and work frequency were also gathered.

Through @replies and retweets, I assessed degrees of interaction and spaces for political conversation, based on the quantity and content of exchanges between Tweeters. The data revealed that users were more likely to retweet/share rather than engage in bidirectional interaction; levels of conversational interaction were low. Predictably, several actors who generated the hashtag #righttofood shared not only an existing interest in food policy issues but also overlapping structural and communicative networks – that is, they were already following one another on Twitter, and also working together on offline projects. For example, Red Cross and the RtF Coalition were co-sponsors, with St Vincent de Paul, St Mary's House of Welcome, Sacred Heart Mission, and the Sydney Food Fairness Alliance, of a special edition of the magazine *Parity* (Council of Homeless Persons, 2016) entitled "Beyond emergency food: Responding to food insecurity and homelessness." This is a good example of how the campaign operated within a wider communication ecology, a constantly evolving set of relationships, processes, and interactions that express the interdependence, mutuality, and co-existence of diverse forms of communication. My geographic analysis of the data revealed the campaign mobilised more than a dozen organisations, ranging from non-government organisations to food relief charities, both nationally and globally. Contributors to the Twitter conversation were concentrated in Sydney and Melbourne but European engagement was generated through FIAN International (the Food First Information and Action Network) based in Germany. The campaign received comparatively significant attention in UK cities including London and Liverpool. The most commonly tweeted and retweeted URL belonged to the Mersey-side based Share Your Lunch organisation which owns the Twitter handle @foodpoverty. The hashtag #foodpoverty was used by several other UK-based food relief enterprises including East London Food Access (@ELFA) which promotes a program called FreshWell that supplies stalls of fresh produce to social housing estates.

Here the most significant finding of the study emerged, namely how hashtags provide a plurality of perspectives on the framing of issues. Commonly used in the UK to describe household food insecurity, food poverty is defined as "insufficient economic access to an adequate quantity and quality of food to maintain a nutritionally satisfactory and socially acceptable diet" (O'Connor et al., 2016: 432). This framing is problematic for advocates insisting on a more nuanced view of the problem that addresses the complexity of food insecurity, which is determined by factors including time, mobility, education, and opportunities to eat socially. Studies interrogating the social construction of food insecurity in high-income nations such as Australia and Canada (SecondBite, 2015; Mah et al., 2014; Rein & Schön, 1996) identify a tendency among policy-makers and program directors to believe that poverty reduction will reduce food insecurity when "while poverty is unquestionable a major contributor to food insecurity, not all the poor are hungry" (Agriculture & Agri-Food Canada, 2002: 59). In affluent economies, food insecurity emerges from a range of factors ranging from lack of cooking skills and limited access to retailers that offer healthy and affordable food to a lack of cooking equipment or basic knowledge of food preparation (Coe, 2013; Food Standards Agency, 2014). Through the sharing of the #foodpoverty hashtag and its sponsorship by social enterprises, city councils, celebrity chefs, and media outlets, it is granted legitimacy as a "catch-all" phrase for food insecurity. As such it simplifies the challenges of people experiencing barriers in accessing healthy, affordable, and culturally appropriate food and stigmatises them as "victims" of food poverty, and passive recipients of food relief.

In summary, the RtF Coalition's Twitter campaign the hashtag #righttofood discursively aligned local and international campaigns and facilitated the cross-flow of information between organisations on multiple scales. It revealed a networked public of pre-existing advocates of the right to food who represent "emergent socio-political assemblages with shared or interlocking concerns who know themselves as, and act as, publics through media and communication" (Burgess & Matamoros-Fernàndez, 2016: 81). As a form of hashtag activism, the campaign supported activists' goals of developing a sense of collective identity. It provides an example of progressive political activism through which elite social media users demonstrate their role in speaking for the food insecure (see also Miller et al., 2016).

The dominance of these representative voices challenges the ideal of social media as de-institutionalised or de-professionalised spaces where "ordinary users and media professionals alike have equally easy access to the digital means of production and distribution of content" (Lomborg, 2011: 65). The construction of narrative forms through social movement processes, including hashtag activism, demonstrates how "agency responds to social conditions and articulates social issues" (Yang, 2016: 16). When a hashtagged word invites a stream of comments and retweets containing personal stories, these posts assume the form of a narrative; therefore "narrative agency is central to hashtag agency" (Yang, 2016). Rhetorical agency is "the capacity to act, that is, to have the competence to speak or write in a way that will be recognized or heeded by others in one's community." It is "communal, social, cooperative, and participatory and, simultaneously, constituted and constrained by the material and symbolic elements of context and culture" (Campbell, 2005: 3). To explore these tensions, more qualitative, adaptive, and reflexive methods are recommended to extend our "contextual and contingent understandings" (Pink, 2009) of digital engagements.

Lessons learned

This study was limited to a study of de-contextualised data captured by TCAT. Interviews and participant observation capturing the "context of utterances" (Bonilla & Rosa, 2015) on the ground would have enabled me to attain a richer account of user motivations, aims, and deliberations in their digital practices and their choices of adopted framings, tactics, and strategies. It might enable me to identify the "connections between online and locality-based realities" (Postill & Pink, 2012: 123) and attain "deeper insights into the movement of messages through networks" (Bonilla & Rosa, 2015: 307). Ideally, these mixed-methods would reveal the degree to which language in hashtags coincides with an intended frame more comprehensively than a close reading or conversational analysis of posts as speech acts. The TCAT word-focused content analysis provides a starting point for interviews with users regarding their communicative intentions. In this manner, the rhetorical role of hashtags as a meta-commentary (Daer et al., 2014) can be explored, revealing how users adapt the hashtag to their "expressive needs" and exercise "vernacular creativity in language use" by, for example, combining hashtags in captions (Utekhin, 2017: 194). This rhetoric function is particularly evident when entire phrases are made into hashtags. Interviews with activists provide an opportunity to explore deliberative, performative functions such as the promotion of solidarity; narration (sharing stories); declaratives (assertions and objections); examples of meta-activism; promotion; and criticism. Findings might inform new framings that resonate with societal understandings of complex issues, as in the case of the #MeToo campaign, which provided informational tools to "promote awareness and educate about sexual misconduct" (Wood, 2018).

To add further value to the study, I would reach out to everyday social media users and particularly those with lived experience of food insecurity who engaged – or chose not to engage – with the campaign. This might help us understand the reticence of ordinary people to engage in self-presentation using stigmatising hashtags such as #foodpoverty and also build foundations for digital inclusion by negotiating the "three-step path" of exploring, building, and participating online and reducing barriers to the creation of local content that currently exist (Surman et al., 2014: 68). Participant-observation and other ethnographic approaches, including unorthodox methods such as photo-voice (Harper et al., 2017), would provide access to voices that are infrequently given the opportunity to speak. Engaging individuals and communities in digital storytelling that focuses on assets rather than deficits in communities, on their preferred platforms, is one way to exploit the emancipative potential of these technologies.

Conclusion

Researching digital food activism entails going beyond the data analytics, such as those generated by the hashtag, to interviewing activists about what they do, how they do it, and why (Lupton & Feldman, 2019; Lewis, 2019; Schneider et al., 2018).

Interviews and participant observation, in combination with analysis of online materials and engagement in campaigns, provide insights into social media activity and sociality spanning online and offline worlds. Data analytics are most useful when complemented by methods that "fill the gap between patterns revealed in big data and meanings that groups of people ascribe to elements of their lifeworlds" (Utekhin, 2017: 187), enabling us to better understand how individuals use various media in their everyday lives. The challenge for digital methods is to take our analysis beyond the most dominant voices to those of previously marginalised stakeholders and facilitate their participation in networked publics, "imagined collective[s] that emerge as a result of the intersection of people, technology and practice" (boyd, 2010: 39). These networks transcend any particular social platform and will be strengthened, for the social media ethnographer, through face-to-face encounters. In these messy, hybrid spaces, researchers, policy-makers, and citizens might engage in new forms of sociality and develop more nuanced and respectful framings of complex issues such as food insecurity.

Dos and don'ts

Do consider how online social network formation also depends on factors external to social platforms

Do consider how data about individual users can harm them if exposed, even when it is publicly available and granted informed consent as per a platform's terms and services

Don't confuse the extent to which you are studying a social issue as opposed to the social media platform itself

Do experiment with new developments, analytics, and visualisation tools such as tweepsmap

Now you try

Find an example of hashtag activism on the platform/s of your choice. Apply Daer's typology to identify how hashtags operate as rhetorical genres in this campaign. Can you find evidence of the power of hashtags to "define" an issue?

Critically reflect on your own practice by analysing your use of hashtags in social media posts. How do you use hashtags to connect, share perceptions, negotiate framings, and implement strategies?

Investigate online tools such as Explore Location to discover how collective identity is displayed and negotiated in your own community, city, or campus.

Experiment with "co-occurrence analysis" by researching the hashtag profile, actor profile, and user profile of a specific hashtag to identify what hashtags appear together, which URLs are connected, what type of user engages with the hashtag, and how these patterns change over time (Marres & Gerlitz, 2016).

Questions for reflection

As discussed in this chapter, hashtag activism on Twitter appears a resource for internal movement collective identity building and organising rather than a means to reach wider audiences. In this case, the community of contributors to #righttofood includes individuals, bloggers, researchers, social enterprises, NGOs, and interest groups concerned about food insecurity around the world. What are the limitations of focusing on a single social media platform in doing this type of research? Hint: consider the medium specificity of different social media platforms.

Ethical considerations

Digital platforms offer researchers unique entry points to complex online and offline worlds, but in our research, we must be aware that the materiality of technology places a limit on the social agency. The functional and relational nature of digital affordances and their capacity to allow or constrain particular kinds of action are important to consider in any analysis of technology in practice (Faraj & Azad, 2012).

Meta fieldwork using online scraping tools that capture large amounts of online data collection may appear unobtrusive but also has ethical dimensions. The researcher, as an external, all-seeing observer, holds an asymmetrical position of power. Accordingly, we must exercise reflexivity and accountability in our research.

Multi-sited digital ethnography provides an opportunity to foreground the voices of people often left out of debates on social issues such as food insecurity. Interviewing vulnerable or marginalised people who have experience trauma or severe and multiple disadvantage requires special consideration on the part of the researcher.

Further readings

Hine, C. (2000). *Virtual Ethnography*. London: Sage.

Pink, S., Horst, H., Postill, J., Hjorth, L., Lewis, T., & Tacchi, J. (2016). *Digital Ethnography: Principles and Practice*. Los Angeles: Sage.

The Digital Methods Initiative. (n.d.). https://wiki.digitalmethods.net/Dmi/DmiAbout

References

Agriculture and Agri-Food Canada. (2002). *Canada's Second Progress Report on Food Security*. Ottowa, Ontario: Agriculture and Agri-Food Canada.

Airoldi, M. (2018). Ethnography and the digital fields of social media. *International Journal of Social Research Methodology*, 21(6): 661–673.

Benkler, Y. (2006). *The Wealth of Nations: How Social Production Transforms Markets and Freedom*. New Haven: Yale University Press.

Bennett, W.L. & Segerberg, A. (2012). The logic of connective action. *Information, Communication & Society*, 15(5): 739–768.

Bennett, W.L. & Toft, A. (2009). Identity, technology, and narratives: Transnational activism and social networks. In A. Chadwick & P.N. Howard (eds.), *The Routledge Handbook of Internet Politics* (pp. 246–260). New York: Routledge.

Bonilla, Y. & Rosa, J. (2015). #Ferguson: Digital protest, hashtag ethnogrphy, and the racial politics of social media in the United States. *American Ethnologist*, 42(2): 4–17. doi: 10.1111/amet.12112

Borra, E. & Rieder, B. (2013). Programmed method: Developing a toolset for capturing and analysing tweets. *ASLIB Journal of Information Management*, 66(3): 262–278.

Borra, E. & Rieder, B. (2014). Programmed method. Developing a toolset for capturing and analyzing tweets. *Aslib Journal of Information Management*, 66(3): 262–278.

boyd, d. (2010). Social network sites as networked publics: Affordances, dynamics and implications. In Z. Papacharissi (ed.), *Networked Self: Identity, Community, and Culture on Social Network Sites* (pp. 39–58). New York: Routledge.

Bruns, A. (2008). *Blogs, Wikipedia, Second Life, and Beyond.* From production to produsage. New York: Peter Lang.

Bruns, A. (2012). How long is a tweet? Mapping dynamic conversation networks on Twitter using Gawk and Gephi. *Information, Communication & Society*, 15(9): 1323–1351.

Bruns, A. & Burgess, J.E. (2011). *The use of Twitter hashtags in the formation of ad hoc publics.* In Proceedings of the 6th European Consortium for Political Research (ECPR) General Conference 2011 (pp. 1–9). Reykjavik: University of Iceland & Brisbane: Queensland University of Technology.

Burgess, J. (2011). The iPhone moment, the Apple brand and the creative consumer: From hackability and usability to cultural generativity. In L. Hjorth, J. Burgess, & I. Richardson (eds.), *Studying Mobile Media* (pp. 28–42). London: Routledge.

Burgess, J. & Matamoros-Fernàndez, A. (2016). Mapping sociocultural controversies across digital media platforms: One week of #gamergate on Twitter, YouTube, and Tumblr. *Communication Research and Practice*, 2(1): 79–96.

Caldwell, M. (2018). Hacking the Food System: Re-Making Technologies of Food Justice. In T. Schneider, K. Eli, C. Dolan, & S. Ulijaszek (eds.), *Digital Food Activism* (pp. 25–42). Abingdon, Oxfordshire: Routledge.

Campbell, K.K. (2005). Agency: Promiscuous and pretean. *Communication and Critical/Cultural Studies*, 2(1): 1–19. doi: 10.1080/1479142040003321 34

Caroll, W. & Hackett, R. (2006). Democratic media theory through the lens of social movement theory. *Media, Culture and Society*, 28(1): 83–104.

Carr, D. (2016, March 26). Hashtag activism and its limits. *New York Times*, viewed January 22, 2016, http://www.nytimes.com/2012/03/26/business/media/hashtag-activism-and-its-limits.html

Chouliaraki, L. (2013). *The Ironic Spectator: Solidarity in the Age of Post-Humanitarianism.* Cambridge: Polity Press.

Coe, S. (2013). Feeding the family, are food prices having an effect?. *Nutrition Bulletin*, 38(3): 332–336.

Couldry, N. (2000). *The Place of Media Power: Pilgrims and Witnesses of the Media Age*, Routledge, London.

Council to Homeless Persons. (2016). Beyond emergency food: Responding to food insecurity and homelessness. *Parity*, 29(2: 48).

Daer, A., Hoffman, R.F., & Goodman, S. (2014). Rhetorical functions of hashtag forms across social media applications. *Communication Design Quarterly*, 3(1): 12–15.

Evans, S.K., Pearce, K.E., Vital, J., & Treem, J.W. (2017). Explicating affordances: A conceptual framework for understanding affordances in communication research. *Journal of Computer-Mediated Communication*, 22: 35–52.

Faraj, S. & Azad, B. (2012). The materiality of technology: An affordance perspective. In P.M. Leonardi, B.A. Nardi, & J. Kallinikos (eds.), *Materiality and Organizing: Social Interaction in a Technological World* (pp. 237–258). Oxford: Oxford University Press.

Food and Agriculture Organisation of the United Nations (FAO) 2004, *Voluntary Guidelines to Support the Progressive Realisation of the Right to Adequate Food in the Context of National Food Security*, viewed September 15 2017, http://www.fao.org/docrep/009/y7937e/y7937e00.htm

Food Standards Agency. (2014). *Food Poverty*, viewed October 7, 2019, https://www.food.gov.uk/business-guidance/food-poverty

Goffman, E. (1959). *The Presentation of Self in Everyday Life*. Garden City, New York: Doubleday.

Goffman, E. (1974). *Frame Analysis: An Essay on the Organisation of Experience*. Cambridge, MA: Harvard University Press.

Harper, K., Sands, C., Horowitz, D.A., Totman, M., Maitin, M., Rosado, J.S., & Alger, N. (2017). Food justice youth development: Using photovoice to study urban school food systems. *Local Environment*, 22(7): 791–808.

Hopke, J.E. (2015). Hashtagging politics: Transnational anti-fracking movement twitter practices. Social Media and Society, July–December: 1–12.

Jensen, J.L., Mortensen, M., & Ørmen, J. (eds.) (2016). *News across Media: Production, Distribution and Consumption*. New York: Routledge.

Kwak, H., Chun, H., & Moon, S. (2011). *Fragile online relationship: A first look at unfollow dynamics in Twitter*. In Proceedings of CHI 2011. 7–12 May. Vancouver, B.C., Canada. viewed January 5, 2016, http://an.kaist.ac.kr/~haewoon/papers/2011-chi-unfollow.pdf

Lewis, T. (2019). *Digital Food: From Paddock to Plate*. London: Bloomsbury.

Lindberg, R., Kleven, S., Barbour, L., Booth, S., & Gallegos, D. (2016). Introduction. In parity. *Beyond Emergency Food: Responding to Food Insecurity and Homelessness*, 29(2): 4.

Lomborg, S. (2011). Social media as communicative genre. *MedieKultur*, 51: 55–71.

Lupton, D. & Feldman, Z. (2019). *Digital Food Cultures*. Abingdon, Oxon.: Routledge.

Mah, C., Hamill, C., Rondeau, K., & McIntyre, L. (2014). A frame-critical policy analysis of Canada's response to the World Food Summit 1988, 2–8. *Archives of Public Health*, 72(41): 1–7.

Maireder, A. & Asserhofer, J. (2014). Political discourses on Twitter: Networking topics, objects and people. In K. Weller, A. Bruns, & J. Burgess (eds.), *Twitter and Society* (pp. 305–318). New York: Peter Lang Publishing Inc.

Marres, N. & Gerlitz, C. (2016). Interface methods: Renegotiating relations between digital social research, STS and sociology. *The Sociological Review*, 64(1): 21–46.

Marres, N. & Weltevrede, E. (2013). Scraping the social: Issues in real-time social research. *Journal of Cultural Economy*, 6(3): 313–335.

Meraz, S & Papacharissi, Z. (2013). Networked gatekeeping and networked farming on# Egypt, *The International Journal of Press Politics*, 18 (2): 138–166.

Meraz, S. & Papacharissi, Z. (2016). Networked framing and gatekeeping. In T. Witschge, C.W. Anderson, D. Domingo, & A. Herminda (eds.), *The Sage Handbook of Digital Journalism* (pp. 95–112). London: Sage.

Miller, D., Costa, E., Haynes, N., McDonald, T., Nicolescu, R., Sinanan, J., Spyer, J., Venkatraman, S., & Wamg, X. (2016) *How the World Changed Social Media*. California: UCL Press.

O'Connor, N., Farag, K., & Baines, R. (2016). What is food poverty? A conceptual framework, *British Food Journal*, 118(1): 429–449.

Ofori-Parku, S.S. &Moscato, D. (2018). Hashtag activism as a form of political action: A qualitative analysis of the #BringBackOurGirls campaign in Nigerian, UK and US Press. *International Journal of Communication*, 12: 2480–2502.

Papacharissi, Z. (2015). *Affective publics: Sentiment, technology, and politics*. New York: Oxford University Press.

Pink, S. (2009). *Doing Sensory Ethnography*. London, Sage.

Postill, J. & Pink, S. (2012). Social media ethnography: The digital researcher in a messy web. *Media International Australia Incorporating Culture and Policy: Quarterly Journal of Media Research and Resources*, 145: 123–134.

Rathnayake, C. & Suthers, D.D. (2018). Twitter issue response hashtags as affordances for momentary connectedness. *Social Media & Society*. doi: 10.1177/2056305118784780

Rauschnabel, P.A., Sheldon, P., & Herzfeldt, E. (2019). What motivates users to hashtag on social media? *Psychology & Marketing*, January, 1–16.

Rein, M. & Schön, D.A. (1996). Frame-critical policy analysis and frame-reflective policy practice. *Knowledge Policy*, 9(8): 5–104.

Right to Food Coalition. (2016). *The Right to Eat Well No Matter Where You Live*, viewed October 20 2016, https://righttofoodcoalition.files.wordpress.com/2016/04/rtf-final_food-access-in-aus_18-april-2016.pdf

Rodak, O. (2019). Hastag hijacking and crowdsourcing transparency: Social media affordances and the governance of farm animal protection. *Agriculture and Human Values*. doi: 10.1007/s10460-019-09984-5

Schneider, T., Eli, K., Dolan, C., & Ulijaszek, S. (2018). *Digital Food Activism*. Abingdon, Oxon.: Routledge.

Steinberg, M. (1998). Tilting the frame: Considerations on collective action framing from a discursive turn. *Theory and Society*, 27(6): 845–872.

Surman, M., Gardner, C., & Ascher, D. (2014). Local content, smartphones and digital inclusion: Will the next billion consumers also be contributors to the Mobile Web? *Innovations: Technology, Governance, Globalisation*, 9(3–4): 67–78.

Turoff, M. & Hiltz, S.R. (2009). The future of professional communities of practice. In C. Weinhardt, S. Luckner, & J. Stößer (eds.), *Designing E-Business Systems. Markets, Services, and Networks. WEB 2008. Lecture Notes in Business Information Processing*, vol 22. Berlin, Heidelberg: Springer. doi: 10.1007/978-3-642-01256-3_13

Utekhin, I. (2017). Small data first: Pictures from Instagram as an ethnographic source. *Russian Journal of Communication*, 9(2): 185–200. doi: 10.1080/19409419.2017.1327328

Weller, K., Bruns, A., & Burgess, J. (2014). *Twitter and Society*. New York: Peter Lang Publishing Inc.

Williams, S. (2015). Digital defence: Black feminists resist violence with hashtag activism. *Feminist Media Studies*, 15(2): 341–344.

Wood, M. (2018). Language in digital activism: Exploring the performative functions of #MeToo Tweets. *Diggit Magazine*. Retrieved from https://www.diggitmagazine.com/academic-papers/language-digital-activism-exploring-performative-functions-metoo-tweets

Yang, G. (2016). Narrative agency in Hashtag Activism: The case of #BlackLivesMatter. *Media and Communication*, 4(4): 13–17. doi: 10.17645/mac.v4i4.692

Yardi, S. & boyd, d. (2010). Dynamic debates: An analysis of group polarization over time on Twitter. *Bulletin of Science, Technology & Society*, 30(5): 316–327.

8

FOOD POLITICS AND THE MEDIA IN DIGITAL TIMES

Researching household practices as forms of digital food activism

Tania Lewis

As the editors of this collection point out, the brave new world of digital food studies brings with it a whole new set of issues related to how we understand, engage with, and research mediated food practices today. Over several decades now, our beliefs, practices, and politics around food have been thoroughly entangled in popular media culture in ways that make them complex objects to study. In digital times, however, how we define and research food "activism" and "politics" *within and in relation to the media space* has become even more difficult. Part of the challenge here is that, in the shift away from "public" forms of broadcast media towards everyday digital engagement, the way food politics is enacted has become much more fragmented and privatised. At the same time, as I argue in my book *Digital Food: From Paddock to Platform* (2020), both food and the digital realm have become intensified sites of public debate and political contestation, with the digital opening up and enabling new possibilities for food activism.

What does this mean for food activism and how media studies scholars might critically engage with and research this area? In the past, we have tended to understand food activists as public figures of some sort, from prominent celebrity chefs such as Jamie Oliver concerned with promoting healthy eating in schools to high-profile food commentators like Michael Pollan. Media studies have also conventionally associated civic activism with newsworthy displays of contestation, such as the anti-globalisation marches that occurred around the world in the 1990s and targeted global food players like Nestlé and Starbucks. Today, digital tools have enabled a range of hitherto hidden actors to come to the fore, from farmers to households and consumers. Digital platforms and social media have enabled these players to connect, organise, and increase the visibility of a range of food issues.

As a media scholar, my own interest has followed this shift over the past decade from a focus on the influential role of celebrity chefs and food TV on food activism to a growing interest in what so-called "ordinary" householders and consumers are doing in relation to shifting critical food practices. In a digital context in which people can minimise food waste via smart phone apps, protest GMO crops through Facebook groups, and where parents can glean huge support for banning chemical food dyes in children's food through online petitions, food activism has become thoroughly entangled in everyday practices, and, for better or worse, enacted on often commercially owned digital platforms.

In this chapter, I examine what these shifts might mean for how we think about and study food-related activism and politics today through and with media. Tracing how the rise of digital engagement has emerged out of and alongside lifestyle media and food television, and discussing how the digital is shaping and enabling various forms of what we might see as food activism and "food citizenship" at an everyday, household level (Booth & Coveney, 2015; Schneider et al., 2018), this chapter reflects on how critical media and food studies scholars might make visible, understand, and research such practices. As a way of investigating new approaches to researching digital food activism, I discuss my own transition from a focus primarily on broadcast media, media texts or programmes, and celebrity chefs on TV, to an increasingly digital ethnographic approach informed by community and household-based food practices in people's daily lives.

From celebrity chefs to everyday digital activists: The changing landscape of food media

Why study food media?

Before we discuss the digital turn in food practices, politics, and activism, I will briefly map out the role of commercial lifestyle media and in particular broadcast television in turning food and culinary culture into a site of everyday politics and ethics as this is key to understanding some of the forms of digital food activism that have since emerged. In recent years, food television, including reality-style and popular documentaries, has been a critical space in which domestic consumers have learned about issues of food justice and insecurity, as well as the problematic practices of agri-business and the fast food industry (Bell et al., 2017; Lewis, 2015; Lewis & Huber, 2015; Phillipov, 2016). Indeed, food programming, and in particular the figure of the celebrity chef, has become one of the main sites where questions around the politics and the ethics of the food we buy, eat, and cook on a daily basis have become mainstreamed (Bell & Hollows, 2011; Goodman et al., 2010; Lewis, 2008).

I have studied the shifting field of food media, food politics, and activism for over a decade. My interest in food media first developed out of broader research into lifestyle media and culture and its growing centrality to our daily lives. This research has been particularly concerned with the growing status and influence of food TV and celebrity chefs, and how they have shaped mainstream audience taste, knowledge, and skills around food and cooking.

Why is studying food media important for food studies?

In my 2008 book *Smart Living: Lifestyle media and popular expertise* I made the argument that in late modern culture, food media has become central to framing food politics. What does this mean about the place of food media in our lives and its importance for understanding food politics more broadly? I would suggest that food politics, as it is portrayed by lifestyle media across print media, broadcast TV, and the digital realm, has become a kind of lifestyle politics. That is, by becoming better eaters, consumers, and even producers of food, people are seen as becoming more responsible and accomplished citizens. The lifestyle gurus and food experts we encounter through lifestyle media, from Paleo diet bloggers to healthy eating gurus such as Jamie Oliver, are thus crucial cultural intermediaries for training people in how to become experts in everyday living (Palmer, 2004), that is, how to be the "best" version of themselves (at least in lifestyle and consumer terms) they can be.

Just as a mediatised culinary culture is now a central part of many people's lifestyles, so too have the politics and ethics of food increasingly entered into the public consciousness. Again, food television and celebrity chefs play an important role here. In the 2000s, for instance, one of the key media spaces where chefs began to emerge as food activists, on television and beyond, was via social documentary formats or what David Bell, Joanne Hollows, and Steven Jones (2017) refer to as the "campaigning culinary documentary" (p. 179). In the UK, Channel Four's Food Fight series – launched in 2008 with *Hugh's Chicken Run* (KEO Films, 2008) followed by *Jamie's Fowl Dinners* (Fresh One Productions and Firefly Film and Television Productions, 2008), a high-profile television event in which Jamie Oliver revealed various unpleasant truths about the industrialised chicken industry – represented a major intervention in this space. Here celebrity chefs such as Hugh Fearnley-Whittingstall and Oliver played a crucial role in giving heightened visibility to questions of food sourcing, animal welfare, and sustainability.

So how and why is this turn to lifestyle experts important for understanding the contemporary landscape of digital food activism? And what has changed in this space as engagements with food media have increasingly shifted from broadcast media to a digital context? What I want to suggest is that ordinary people (rather than just celebrities) have increasingly become agents of change within and through food media. At the same time, we have seen a shift from food media being driven by traditional media institutions to the growing role of social media and participatory digital platforms. Crucial in the Food Fight series, for instance, was the growing emphasis not just on chefs as political actors but on *consumers* as potential everyday activists. For example, in 2010, Fearnley-Whittingstall launched a major Fish Fight campaign against the then European practice and policy of discarding up to 50%

of dead edible fish found in fish catches (River Cottage, n.d.). Targeting consumers initially in the UK, the website featured the contact details and fish sourcing policies of the major UK supermarkets with the following call to arms:

> The power to protect our oceans is in your hands now, so make sure our supermarkets and our politicians hear your voice. Ask them a question, tell them what you think, and don't give up until they respond to you!

The Fish Fight campaign soon spread to other parts of Europe with celebrity chefs and other spokespeople using a mixture of online, TV, radio, and press and social media. For example, an online tool allowed "Fish Fighters" everywhere to contact their local politicians in their native language. Due in large part to people power, 3 years later the campaign saw an EU vote against the practice and a major change of European fisheries policy.

Social media played an important role in the success of the Fish Fight campaign. This example thus provides a useful segue from the space of broadcast food television to the digital realm where a number of websites, tools, and technologies have proliferated in recent years which target the ethical food consumer. If ethically and environmentally minded celebrity chefs have sought to inform consumers about how their everyday actions in the supermarket and kitchen might impact on local producers, supermarkets' sourcing policies, animal welfare, and sustainable fisheries, the digital realm appears to offer householders the opportunity to take food citizenship to the next level. Through providing the consumer-citizen with a range of tools to link their personal and household food routines and habits to issues of responsible and ethical food consumption, the digital realm shifts food media from a focus on *representing* food issues and *informing* consumers to enabling potential *practices* of engagement. Through social media, ordinary members of the public are at least potentially able to engage in everyday forms of food activism without necessarily taking part in a street march or darkening the doors of a court room.

Digital tools and platforms: The rise of food apptivism and alternative food networks

One example of these *potential* practices of engagement is seen in the growing use of applications (apps) intended to support ethical food consumption or what has been termed more broadly "apptivism." In their systematic research on mobile ethical consumption apps, Humphery and Jordan (2016) argue there are three types on the market. The first are largely informational apps (such as The Good Shopping Guide and Shop Ethical!) while the second are apps that reply on barcode technology. For instance, The Good Guide allows you to use barcode technology at the supermarket to get just-in-time information about the environmental, health, and social impacts of companies and products. The third group of apps they discuss are those such as the Buycott app that employ crowdsourcing technology to collate relevant information and to also potentially connect with a network of activists. One example of this is anti-palm oil activists.

There are also a range of apps that enable users to move beyond the realm of shopping. For instance, Australian website Spare Harvest is concerned with connecting local farmers, gardeners, and cooks around the world who want to swap, sell, or share their excess harvest through a global map where people can click onto local postings. As the founder puts it: "It's about making sure that all those valuable resources [...] don't end up in landfill" (Nichols, 2017). Similarly, OLIO, an app aimed at reducing food waste, enables neighbours to connect with each other and/or with local businesses to exchange any surplus food they might have. OLIO uses can also become Food Waste Heroes, which involves collecting unsold food from local food businesses, bringing the food home, listing it on the OLIO app, and redistributing to their neighbours. OLIO thus emphasises civic-minded digital engagement with a commercial edge.

The digital food space has also seen the emergence of interactive platform technology that aims to connect ethically minded consumers to a range of actors and groups concerned with offering alternatives to global agri-business. In an article entitled "Virtual Reconnection: The online spaces of alternative food networks in England," Elizabeth Bos and Luke Owen (2016) explore how the digital realm offers opportunities for reconnection with the "complex systems of food provisioning" that have worked "to distance and disconnect consumers from the people and places involved in contemporary food production" (p. 1). As their research on online food hubs suggests, digital networks are as much about strengthening connections to place and the local as they are about forging global links, often involving complex co-articulations of practices and connections between consumers and producers occurring via both offline and online spaces and practices.

Changing media methods for changing times

Given these changes, how might media studies researchers study digital food activism today? What sorts of tools, research design, and approaches does one need to take? As we've seen, the food media landscape has shifted from a broadcast modality embedded primarily in media institutions like TV, in audience research, and in a focus on media texts and meanings, to one that is increasingly embedded in the actions and practices of householders and consumers. Simultaneously, the field of media studies more broadly has seen a recent turn towards embracing people's "social practices." Indeed, media scholar Nick Couldry (2010) has called for nothing less than a new media paradigm that "sees media not as text or production economy, but first and foremost as practice" (p. 35). In recognizing the need for media studies to move from a sole focus on meanings, texts, and institutions, David Morley (2007), another key British media and cultural studies scholar, has called for a "non-mediacentric media studies" in order "to better understand the ways in which media processes and everyday life are interwoven with each other" (p. 200).

The question for media researchers interested in digital food activism then becomes "What, quite simply, are people *doing* in relation to media across a whole range of situations and contexts?" (Couldry, 2010: 39). Here it should be noted that I am not suggesting that texts and discourses exist in a realm separate from embodied actions and practices. They are of course thoroughly intertwined; practices are discursively shaped while texts and discourses have material, social, and historical impacts in the world. Nevertheless, media studies have, until relatively recently, tended to separate out research on texts from that on audience or industry practices; hence the "turn" towards a non-mediacentric media studies.

However, the practice turn in media studies has left a methodological gap. Mobile phones, laptops and tablets have become so integral to daily life and to our interpersonal and broader social relationships that it has become increasingly impossible to separate out the world of digital media practices from the rhythms and routines of everyday living. A key difficulty new media researchers face is thus precisely the ubiquity and everydayness of the digital itself. How do we research digital food politics when "activist" engagement has likewise become embedded in the digital routines and habits of our everyday lives in ways that are both central to daily existence but also increasingly taken for granted and invisible?

There are many different ways to research the shifting and emergent field of digital food activism depending on the type of activism one is examining. One important critical approach which nicely complements a non-media centric media studies and that I focus on here is digital ethnography. In our collectively co-authored book *Digital Ethnography: Principles and Practice* (Pink et al., 2016), we argue for "a particular type of digital ethnography practice that takes as its starting point the idea that digital media and technologies are part of the everyday and more spectacular worlds that people inhabit" (p. 7). Just as the space of food politics thoroughly effaces the line between public and private engagement and activism, so too does digital ethnography see the digital (and thus digital activism) as both eventful and everyday, and as linked to *practices* or the ways in which we engage with the world, through "everyday life" or "lifeworlds" (de Certeau, 1984; Husserl, 1966).

Researching digital food activism as *practices* can include focusing on a wide range of activities from people's uses of ethical apps in their daily lives to the use of Facebook-enabled alternative food movements to mobilise political engagement. Here research design might include a range of different "sites," from online ethnography to attending public protests organised online against GMO food. Depending on the type food activism, one's methods are likely to be similarly diverse; they might involve collecting (or "scraping") digital data from platforms, and interviewing and/or "shadowing" users as they engage with apps and/or fieldwork with key intermediaries in food movements.

Digital ethnography is thus not a fixed method or approach but is tailored to different sites and practices, something that is the case ideally for all empirical research. As we argue in our book, a keyword here is "multiplicity," that is, there is more than one way to engage with the digital: "digital ethnography research is always unique to the research question and challenges to which it is responding" (Pink et al., 2016:

p. 8). Adopting a digital ethnographic approach involves recognising that there is no single methodology that can capture the field of digital food activism but rather researchers need to be flexible and open.

As a way of examining some of the new methodological approaches and tools required by media researchers in a digital food context, in the next section I discuss the digital ethnographic research I conducted with members of Melbourne Permablitz, an online permaculture community (Permablitz, n.d.). Melbourne Permablitz is an ad hoc network of people who volunteer to transform suburban gardens into productive food systems, designed according to the principles of permaculture.[1]

Doing Digital Food Ethnography: Researching household permaculture as digital food activism

> *I arrived somewhat late in the morning to the Sunday 'blitz', driving up through a part of northern Melbourne I hadn't visited before. Armed with a video-camera, shovel, hat and sunscreen I followed a lanky stranger down the side drive of an ordinary suburban brick house to find a good sized group of people already at work weeding, hacking away at plants and thoughtfully inspecting the various spaces and 'projects' underway in the, to my (inner urban) eyes, rather huge quarter acre block.*
>
> *Cut to the end of the day and I and others, no longer strangers, are taking photos of (and in my case videoing) the transformation that has taken place during the day. An old chicken coop has been repaired and extended, once desolate patches of dying off lawn turned into wooden-framed raised garden beds, overgrown spaces cleared and turned into potentially productive land ready for planting .... In one day, with the aid of planning and the labour and skills of many bodies, a large neglected suburban backyard is on its way to turning into an integrated permaculture garden complete with chickens.*

As this quote from my fieldnotes or reflections on my experiences in the field suggests, my research with Melbourne Permablitz primarily involved participatory fieldwork in suburban gardens. Over a 1-year period, I attended and participated in a number of blitzes at a range of suburban sites across Melbourne. At each backyard blitz (or permaculture "makeover") I attended, I worked for a day alongside fellow Permablitzers, I talked to volunteers, home owners, and blitz organisers (blitzes are extensively planned and led by volunteers who usually have permaculture training). At the same time I took field notes (fieldnotes are a keystone of ethnography and involve everything from jotting down sensory perceptions and personal observations to capturing questions and critical reflections one has in the field) as well as photos and videos of blitz activities and the dramatic transformational process undertaken at backyard sites using a mix of hand-held devices (video camera and iPhone) depending on the circumstances. I then shared images and videos captured with the householder in question for posting on the Permablitz website (alongside other "blitzers" who also documented the process for the website). Thus, while I was in part interested in studying the role of the Permablitz website *itself* in enabling strangers from across suburban Melbourne to connect and work together to convert traditional Australian backyards into sites of productive food growing, I was also interested in *using* digital media to research and foreground the role of seemingly privatized household sustainability initiatives *as forms of digitally enabled food activism*. Although, as I've noted, my own use of digital documentation mimicked that of the community itself (such as using videos and "before and after" shots to recruit people to the permaculture cause), my research was not on people's media practices *per se* but rather on seeing these digital engagements as part of a set of broader forms of civic engagement or activism around urban farming and sustainable food (Figure 8.1).

This research approach was markedly different from my previous research on food media and celebrity chefs, which involved a mixture of analysing media texts and interviewing TV producers and sometimes audience members in their homes. As the fieldnote that opened this section indicates, a central aspect of my research on food gardening practices and the digital has been to actively participate in and acquire skills around those same practices. My concern with both making visible and actively contributing to processes of change, in this case the transformation of a domestic garden into a sustainable food space, can be seen as an extension of the change-making premises of "action research." Simultaneously, my positioning as a co-participant draws upon elements of "experiential ethnography" in terms of my focus on gaining knowledge about the "culture" of digital food practices through processes of intense immersion (Sands, 1999; Sparkes, 2009).

The ethnographic researcher therefore takes on a very different positionality from the traditional media studies scholar who interprets texts and/or observes and interviews media production staff and audience members. Through adopting a

FIGURE 8.1 A Permablitz in a suburban Melbourne garden where the lawn was transformed into raised beds for growing vegetables.

digital ethnography approach, I was located as an active co-participant (albeit one who is institutionally based and not completely an insider within the Permablitz community). This is not to say that ethnography cannot be employed in researching food TV and/or the practices of celebrity chefs. Of course it can and *does* get used in these settings, but in my own research, it was my experience of grassroots food and sustainability groups that encouraged me to shift my own research interests from one primarily concerned with texts and programmes to one focused on household and community actions. Another aspect of digital ethnography which distinguishes it from conventional media research is its focus on embodiment, communicating somatic sensory practices and the visceral nature of the everyday (Hayes-Conroy & Martin, 2010) in "nonlanguaged ways," as Panhofer and Payne (2011), writing about researching dance, put it. Research on food media and activism for instance often fails to account for the embodied and connective pleasures of engaging in the food space, whether that involves urban gardening with community to grow vegetables, sharing meals and food experiences via apps and platforms, or connecting with one's local farmer through direct-to-consumer produce. Through the use of mobile technologies such as video cameras and mobile phones, combined with ethnographic research on the sensory and haptic dimensions of everyday practices, digital ethnography enables a complex engagement with the sights, sounds, taste, smell, and feel, the rhythms and temporalities of a range of actors, spaces, and practices (Pink, 2009).

Methodologically, the use of mobile visual technologies in combination, with the moving, labouring body of the researcher, can also be seen as extending notions of empirical research. If empiricism often privileges the beliefs and actions of humans, digital ethnography moves us beyond that of purely visible and/or human, behavioural markers of social change towards what Clough (2009) terms "infraempiricism." While media studies are often human centred, my video-based research captured the role of non-human "actors" in the food activism, such as chickens, soil, gardening tools, and other environmental and material elements (Strengers & Maller, 2012). This "more-than-human" approach dovetails with a growing public consciousness around the essential role of animals and environment in our foodscapes. This awareness has, of course, been fuelled in part by more traditional media such as popular documentaries like *Food Inc* (Kenner & Kanopy, 2014) and *Fast Food Nation: Do you want lies with that?* (McLaren & Thomas, 2006), and Michael Pollan's (2007) best seller *The Omnivore's Dilemma: A natural history of four meals*.

My research on Permablitz Melbourne also foregrounds the role of the researcher and the video camera and/or smart phone in *constructing* the space and place of the research encounter. Broadcast media frame certain activities, practices, and people as events and actors worthy of attention (for instance, celebrity chefs and more recently farmers like Joel Salatin). Through visually documenting suburban productive gardening as a form of "food activism," my research likewise turns otherwise hidden activities such as backyard makeovers into "events." The very act of videoing and publicly posting images on an urban agriculture website of a suburban backyard that has been converted into a productive garden involves intervening in and transforming everyday practices, such as gardening and planting food crops, into forms of activism. As Pink (2009) argues, "ethnographic uses of audiovisual media can be understood as both a research technique and as practices that become co-constituent of an ethnographic place" (p. 101).

Learning from the ethnographic turn

What then are the potentials and challenges for media researchers in a digital food studies context? What lessons might be drawn from the research discussed above? One *potential* that the above research highlights is the ability of ethnography to foreground the role of practices that engage the digital in unexpected ways. The shift from the methods associated with broadcast food media and their focus on text/industry/audiences moves us into a very different space for conceptualising "mediatised" food activism as a set of social practices tied to and enabled by (but not centrally focused on) the digital. In the case of Permablitz, for example, while the "community" only exists as such through a collaborative website, ethnographic research on the actual backyard permaculture practices enacted by the volunteers provides a more holistic picture of that community and its activities than an analysis of the website would provide. It also shows the ways in which digital media practices are intertwined with a range of other sites, practices, technologies, and actors – in this case, gardens, chain saws, hammers, and smart phones; volunteers who both labour and record the blitz in action; and the worms, chickens, and other non-human actors involved in "making" a permaculture space. The research discussed above thus illustrates the *non-digital-centric-ness* of digital ethnography, that is, how this kind of research can encompass "everyday life activities or localities that are not usually contexts or sites of digital media immersion, or are sites of limited digital media immersion or availability" (Pink et al., 2016: 10).

But what might be the challenges of such a de-centred approach to media and activism in a digital context? A major difficulty is how one defines the boundaries of the sites and practices to be researched. If, as Wendell Berry (2009) has said, "eating is an agricultural act," there is a sense that researching food activism today might be a similarly boundless task in which we must necessarily engage with all the wider social, economic, and geographic implications of whatever digitally enabled food activism we are researching. Clearly, it is not possible for media researchers to individually take on such a remit, but such a view does suggest the need for significant reflexivity and transparency about our research design and how we define research sites, actors, and practices. For instance, if we are researching the use of social media for organising street-based GMO food protests (as per Adamoli's (2012) research on the role of Facebook in driving the Right to Know rallies in the United States), do we need to also conduct fieldwork with offline protesters as well?[2] If we are concerned with making the role of the digital in enabling food activism visible, what are the commercial processes of automation that shape these communicative practices? Do we need also to understand the algorithmic processes and practices underpinning social media?

The complex, multi-site nature of digital food practices and their embeddedness in everyday practices also speaks to another challenge and potential – the role of collaborative and participatory models of research design. In conducting my research with the Permablitz community, I was aware of some key limitations or questions around a solo researcher "dropping in" to the Permablitz community and videoing and analysing the actions of this group of people. For instance, I was conscious of the limits of studying practices that are centered on action and transformation by attending and participating in one-day blitzes over the period of a year. At times I felt uncomfortable about my ambiguous status at Permablitzes and the ethics of using the material I had gleaned as "research." Was I helping or hindering the change agenda so central to the Permablitz community? What were the ethics of participating in and making visible household food practices in which people had invested considerable amounts of time and expertise?

How to do digital food activism research?

Given these reflections, how might media studies scholars conduct research that is robust but also ethical? How do we conduct research that reflects the increasingly embedded and everyday nature of digital engagement while also understanding the ethical challenges that come with researching people's every day, and the often personal and/or domestically based practices that come along with this? How might our research account for the fact that the people we "study" often have what I've termed elsewhere "ordinary expertise," that is, considerable skills and knowledge in relation to everyday life and digital practices? What then are some of the "dos and don'ts" for conducting research in the digital food space today?

Do work collaboratively with your informants to co-design the research.

Don't assume that you know what you're going to find in the "field" (or, indeed, that you know what the "field" actually is).

Do recognise that people are data collectors and data sharers. You can work collaboratively with them to share research materials in ways that enrich their lives and that assist your research in the process.

Don't use people's personal experiences, data, and labour without discussing this with them and talking about issues of recompense.

Do take an open, ethnographic approach to the role of technology in digital food activism. Be prepared to be surprised!

Don't try and pave over the complexities of your findings by, as Eva Giraud (2018) puts it, adopting a predetermined stance on digital affordances that neglects "the messiness of activists" material engagements with digital media, and all the frictions and unexpected manifestations of agency that go with it' (p. 140).

Activity box

Design a research project for studying food waste practices and the use of digital technology. For instance, you might want to choose to focus on a food business and/or cafe that has signed up to an online platform aimed at waste reduction and at distributing food to households in need. What research question/s would you ask? What kinds of approaches and methods would be best for answering the specific questions you have? What concepts or theories might be useful for framing the concerns of your project? What kind of ethical issues and/or difficulties might the project face? Is this a project you would conduct as a solo researcher or as part of a team? Would you involve the community and cafe itself in the research? If yes, how and why?

Questions for reflection

In the area of food activism, the people involved are often politically and digitally active and may also be conducting their own research as part of their activist practices. How do we define research and the role of the academic researcher in such contexts? How might we rethink terms like "informant" and research "subject"?

Studies that use big data tend to get a lot of media attention. What might be the limitations of this kind of quantitative approach to researching digital food activism? What are the benefits of the kinds of qualitative research we have discussed in this chapter?

With many households and communities increasingly involved in trying to live more sustainably and ethically, in what ways might we understand contemporary food activism? What role might the digital play in enabling these kinds of lifestyle practices?

Further readings

Lupton, D. (2018) Cooking, eating, uploading: Digital food cultures. In LeBesco, K. & Naccarato, P. (eds.), *The handbook of food and popular culture* (pp. 66–79). London: Bloomsbury.

Middha, B. (2018) Everyday digital engagements: Using food selfies on Facebook to explore eating practices. *Communication Research and Practice* 4(3), 291–306.

Originally Australian (now global) online collective and software platform aimed at developed an alternative online food system. (n.d.). Retrieved from https://openfoodnetwork.org.au/

App aimed at reducing food waste. (n.d.). Retrieved from https://olioex.com/about/

Notes

1 The idea of permaculture was developed in the mid 1970s by Australians Bill Mollison and David Holmgren (1978) as an alternative to industrialised forms of agriculture (See also Holmgren, 2002; Mollison, 1988). Conceived of as an ethical and holistic design system for sustainable living, land use, and land repair, '[p]ermaculture has come to mean a design system, for taking patterns and relationships observed in natural ecosystems into

novel productive systems for meeting human needs' and has been embraced by individuals, groups, and communities worldwide (Permablitz, n.d.).

2 See Adamoli's (2012) research on the role of collectivised forms of online civic engagement in food politics, from grassroots permaculture movements to social media driven protests such as The 'Right to Know Rally' of 2011 in which activists demanded that the US government introduce GMOs food labelling across the country.

References

Adamoli, G. (2012). *Social media and social movements: A critical analysis of audience's use of Facebook to advocate food activism offline.* (Doctor of Philosophy), Florida State University, ProQuest Dissertations Publishing.

Bell, D., & Hollows, J. (2011). From river cottage to chicken run: Hugh Fearnley-Whittingstall and the class politics of ethical consumption. *Celebrity Studies*, 2(2), 178–191. doi: 10.1080/19392397.2011.574861

Bell, D., Hollows, J., & Jones, S. (2017). Campaigning culinary documentaries and the responsibilization of food crises. *Geoforum*, 84, 179–187. doi: 10.1016/j.geoforum.2015.03.014

Berry, W. (2009). Wendell Berry: The pleasures of eating. *Centre for Ecoliteracy*. Retrieved from https://www.ecoliteracy.org/article/wendell-berry-pleasures-eating

Booth, S., & Coveney, J. (2015). *Food democracy: From consumer to food citizen.* Singapore: Springer.

Bos, E., & Owen, L. (2016). Virtual reconnection: The online spaces of alternative food networks in England. *Journal of Rural Studies*, 45, 1–14. doi: 10.1016/j.jrurstud.2016.02.016

Clough, P. (2009). The new empiricism: Affect and sociological method. *European Journal of Social Theory*, 12, 43–61.

Couldry, N. (2010). Theorising media as practice. In Postill, J. & Bräuchler, B. (eds.), *Theorising media and practice* (pp. 35–54). New York: Bergahn Books.

de Certeau, M. (1984). *The practice of everyday life.* Berkeley: University of California Press.

Fresh One Productions & Firefly Film and Television Productions. (2008). *Jamie's fowl dinners* [Television movie]. United Kingdom: Channel 4.

Giraud, E. (2018). Displacement, 'failure', and friction: Tactical interventions in the communication ecologies of anti-capitalist food activism. In Schneider, T., Eli, K., Dolan, C., & Ulijaszek, S. (eds.), *Digital food activism* (pp. 130–150). Retrieved from https://ebookcentral.proquest.com

Goodman, M. K., Maye, D., & Holloway, L. (2010). Ethical foodscapes?: Premises, promises, and possibilities. *Environment and Planning A*, 42(8), 1782–1796.

Hayes-Conroy, A., & Martin, D. G. (2010). Mobilising bodies: Visceral identification in the Slow Food movement. *Transactions of the Institute of British Geographers*, 35, 269–281.

Holmgren, D. (2002). *Permaculture: Principles and pathways beyond sustainability.* Hepburn: Holmgren Design Services.

Humphery, K., & Jordan, T. (2016). Mobile moralities: Ethical consumption in the digital realm. *Journal of Consumer Culture*, 18(4), 520–538.

Husserl, E. (1966). *The phenomenology of internal time-consciousness.* Bloomington: Indiana University Press.

Kenner, R., & Kanopy (2014). *Food Inc.* United States of America: Kanopy Streaming.

KEO Films. (2008). *Hugh's chicken run* [Television series]. United Kingdom: Channel 4.

Lewis, T. (2008). *Smart living: Lifestyle media and popular expertise.* New York: Peter Lang.

Lewis, T. (2015). 'One city block at a time': Researching and cultivating green transformations. *International journal of cultural studies*, 18(3), 347–363. doi: 10.1177/1367877913513694

Lewis, T. (2020). *Digital food: From paddock to platform*. London: Bloomsbury.
Lewis, T., & Huber, A. (2015). A revolution in an eggcup?: Supermarket wars, celebrity chefs and ethical consumption. *Food, Culture & Society*, 18(2), 289–307. doi: 10.2752/175174415X14190821960798
McLaren, M., (Producer), Thomas, J., (Producer). (2006). *Fast food nation: Do you want lies with that?* (Widescreen ed). Australia: Magna Pacific.
Mollison, B. (1988). *Permaculture: A designer's manual*. Tyalgum: Tagari Publications.
Mollison, B., & Holmgren, D. (1978). *Permaculture one: A perennial agriculture for human settlements*. Tyalgum: Tagari Publications.
Morley, D. (2007). *Media, modernity and technology: The geography of the new*. London: Routledge.
Nichols, J. (2017). Spare Harvest battles food waste by swapping, selling and sharing excess produce. *Australian Broadcasting Corporation – ABC Rural*. Retrieved from https://www.abc.net.au/news/rural/2017-11-06/spare-harvest-website-stops-food-waste/9115932
Palmer, G. (2004). 'The new you': Class and transformation in lifestyle television. In Holmes, S. & Jermyn, D. (eds.), *Understanding reality television* (pp. 173–190). London: Routledge.
Panhofer, H., & Payne, H. (2011). Languaging the embodied experience. *Body, movement and dance in psychotherapy*, 6, 215–232.
Permablitz. (n.d.). About Permablitz. *Permablitz: Eating the Suburbs – One Backyard at a Time*. Retrieved from http://www.permablitz.net/about-permablitz/
Phillipov, M. (2016). The new politics of food: Television and the media/food industries. *Media International Australia*, 158(1), 90–98.
Pink, S. (2009). *Doing sensory ethnography*. London: Sage Publications Ltd.
Pink, S., Horst, H., Postill, J., Hjorth, L., Lewis, T., & Tacchi, J. (2016). *Digital ethnography: Principles and practice*. London: Sage.
Pollan, M. (2007). *The omnivore's dilemma : A natural history of four meals*. New York: Penguin.
River Cottage. (n.d.). Hugh's fish fight. *River Cottage*. Retrieved from https://www.rivercottage.net/campaigns/hughs-fish-fight
Sands, R. R. (1999). *Experiential ethnography: Playing with the boys. Anthropology, Sport, and Culture*. Westport: Bergin & Garvey.
Schneider, T., Eli, K., Dolan, C., & Ulijaszek, S. J. (2018). *Digital food activism* (1st ed.). Abingdon: Routledge.
Sparkes, A. C. (2009). Ethnography and the senses: Challenges and possibilities. *Qualitative Research in Sport and Exercise*, 1, 21–35.
Strengers, Y., & Maller, C. (2012). Materialising energy and water resources in everyday practices: Insights for securing supply systems. *Global Environmental Change*, 22(3), 754–763.

9

EXPLORING THE FOOD BLOGOSPHERE

Meghan Lynch and Kerry Chamberlain

This chapter focuses on collecting and analysing data from food blogs for qualitative food studies research. We first discuss food blogs and food blogging communities. We then comment on some methodological approaches for researching food blogs, with a focus on the methodology we have used in food blog studies, *netnography*. Netnography involves becoming immersed in and understanding the online posts of food bloggers. Next, using examples from our own and other food blog research, we highlight the potentials and challenges in researching food blogs. Finally, we offer three considerations for researchers interested in studying food blogs. This chapter provides researchers with an overview of food blogs, real-life examples of research practices used in food blog research, and recommendations for how to conduct research on food blogs.

Food blogs and food blogging communities

Blogs have been described as a lens into understanding how people are transforming what was previously considered personal information into public property (Gurak & Antonijevic, 2008). Unlike a personal webpage, a blog is updated on a regular basis, many almost daily, making it rather like an online diary (Gunter, 2009; Hookway & Snee, 2017). Although blogs may share many characteristics of diaries, they possess important distinctions. Blogs are both *interactive*, as many have loyal followers who leave comments and engage in discussions on blog posts, and *intertextual*, in that they link to other blogs, through permalinks (links to specific postings), trackbacks (identifying the creator of a blog post link), and blogrolls (promoting other blogs of related interest for readers to follow). This intertextuality also enhances possibilities for interaction among bloggers (Li, 2007).

The online context has been found to encourage blog authors (bloggers), to reveal more information, including personal information, than traditional research methods, such as interviewing or focus groups (Hookway & Snee, 2017). de Laat (2008), exploring why bloggers revealed such intimate details of their lives, found that, as a consequence of reading and commenting on one another's blogs, blogging communities developed a sympathetic climate of trust. Blogging has also been found to encourage social support and a sense of belonging, and blogging communities can function as security spaces (Baker & Moore, 2008). Indeed, seeking emotional support has also played an important role in bloggers' intentions in creating blogs (Miller & Sheperd, 2004). For these reasons, blogs provide an innovative data source for researchers. Further, blog entries are posted in a public arena, allowing for unobtrusive access to narratives of bloggers' experiences and perspectives (Hookway & Snee, 2017).

In this chapter, we examine research on, and discuss how to research, food blogs as sites for analysis and interpretation. Blog research may consider a single blog (e.g., Howarth, 2017) or consider material drawn from across a range of related blogs, all focused on the researcher's topic of interest (e.g., Véron, 2016). In the latter case, we can consider the related set of food blogs as a community of interest (in Véron's research, the community of interest was French vegan bloggers).

Food, in one form or another, is a popular topic for blogs. It is almost impossible to estimate how many food blogs exist, but there were considered to be more than 16,000 in 2012 (Lofgren, 2013) and numbers have increased substantially since then. The popularity of food blogging has been attributed to the increasing societal interest in food and food-related matters, alongside the increased ease of use of blogging technology, and the popularity of telling personal stories and creating communities of interest (Lofgren, 2013; Morrison, 2011; Salvio, 2012). Although more likely to be authored by people with higher levels of education (Schradie, 2012), and largely the province of younger (25–45 years old) women (Norén, 2011), large numbers of food blogs are written by "ordinary" people, so-called hobby bloggers. Lofgren (2013) has noted the movement of some food bloggers from a hobbyist activity to a cottage industry, where blogging leads to publishing cookbooks, writing food memoirs, or appearing on other food-related media, such as television. Some food blogs are focused entirely on commercial interests and the generation of revenue from marketing food-related products (e.g., Lepkowska-White & Kortright, 2018). Food blogs can be quite varied in their focus. For example, Sanford (2010) explored how weight-loss-focused food blogs can serve as support tools, while Rodney (2018) investigated the discourses of healthy eating in healthy living blogs. Hart (2018) critically analyzed vegan food blogs written by men to examine the relationship between gender and veganism. Salvio (2012), analyzing "culinary autobiographies" – food blogs that interweave personal narratives with food recipes, images, and recommendations for cooking and eating – documented the challenges of balancing domesticity and everyday living. Howarth (2017) discusses a case study of an austerity food blogger – a women struggling to feed her family while on welfare – to illustrate how blogs can politicise social issues. Similarly, Elliott et al. (2017) provide a narrative analysis of two blogs written by food-insecure mothers to document the normative and transgressive narratives presented around mothering, family, and resource constraints.

Our own research with food bloggers has added to this literature. Lynch (2010, 2012) examined the blog content and virtual socializations in a community of self-described "healthy living food blogs." This revealed that this food-blogging community has an established set of conventions for both acceptable blogging and dietary practices, with bloggers discussing how much support and dietary information they received from the community. These findings were troubling in view of the analysis that the food bloggers promoted quite specific attitudes and behaviors, and the conventions and themes of this blog community can be seen to illustrate many of the negative behavioral and psychosocial associations of dietary restraint.

Leggatt-Cook and Chamberlain (2012) explored blogs that were focused on weight-loss, finding that a primary motivation for weight-loss bloggers was to make themselves personally accountable to an audience. To do so, they had to construct a version of themselves in the public domain and represent their changing bodies though the use of before-and-after images and stories of food and eating, success, and failure. The weight-loss blogosphere therefore was a dynamic and ambiguous space where bloggers had to negotiate competing discourses around fatness. Their writing and representational styles create a community, where bloggers rely on their readers for accountability and support in achieving their goals, and where readers receive inspiration and involvement. Leggatt-Cook and Chamberlain concluded that blogging has become a technology for achieving weight loss.

How can researchers study food blogs?

In this chapter, we focus on the qualitative analyses of blogs, but note there are a range of quantitative approaches possible, such as content analysis (which involves some type of numerical description, for example, counting the number of times certain blog features appear in your sample) and network analysis (which uses tools to measure, calculate, and visualize related aspects of blog communities) (Hookway & Snee, 2017). There are a wide range of applicable qualitative methodologies for researching blogs, including narrative and discursive approaches. Many blogs also contain photographs and images that can be incorporated into the analysis (Mannay et al., 2019). Food-related blogs tend to include many images, and including these in the analysis may be highly valuable for answering your research question (see e.g., Taylor & Keating, 2018).

For researchers interested in in-depth interpretative analyses of food blogs, we recommend the research methodology known as *netnography* (also known as "cyber-ethnography," "webnography," and "virtual ethnography"), a methodology that involves adapting ethnographic practice and includes such methods as textual analysis (Fairclough, 2003), to the online world. Ethnography is a qualitative research methodology whereby a researcher immerses herself in the everyday life of a person or a community with the goal of understanding life from that person's or community members' perspectives; netnography maintains these principles but in an online setting (Kozinets, 2020) and can be considered as subsumed under the more general term of digital ethnography (Pink, 2016). Netnography can involve researchers passively observing social media discussions (passive netnography) and/or using social media actively to involve participants in research (active netnography) (Kozinets, 2020). We will focus on passive netnography (referred to as "netnography" for the remainder of this chapter). Although netnographic research draws on general principles of participation and interaction, is naturalistic, and is unobtrusive, there is variation in netnographic research as the "approach to individual research questions, sites and forms of understanding is idiosyncratic" (Kozinets, 2017: 375). In researching food blogs, we emphasize the naturalistic and unobtrusive aspects and minimize the participatory and interactive aspects.

Netnography has several advantages for researching blogs and blogging. This method provides an unobtrusive process of observing people's more naturally occurring discussions in a context that has not been created by the researcher (Kozinets, 2020); it passes the dead researcher test (Potter, 2002), in that the data would have been generated whether or not the researcher was alive at the time. Also, researchers can obtain an extensive body of data quickly because they can gather effectively transcribed material from blog posts and their accompanying online discussions (Kozinets, 2017, 2020). Netnographers can collect data across considerable time frames, months or even years, depending on how much is required for the research aims and can be managed in analysis. As with any form of social research methodology, netnography offers advantages and disadvantages for researching, but its use over a wide range of research topics and settings serves to indicate that it has considerable value for researching online communities, including blog and blogging research.

Working in the field

In this section, we discuss our experiences in researching two different types of food blogging communities: first, healthy living blogs and second, weight-loss blogs.

Researching healthy living blogs

My food blog study (Lynch, 2010, 2012) was inspired by research on the importance of social support for establishing and maintaining healthy dietary behaviors (Strong et al., 2008). I wondered: what could food blogs offer in terms of social support for healthy eating? With this focus, I wanted to locate a community of food bloggers that self-described as promoting healthy dietary habits. Finding such a community proved surprisingly easy. I first used Google, applying the Google Blog Search (unfortunately, this filter was discontinued in 2011), searching for blogs that included terms associated with healthy eating, such as "healthy," "fit," and "nutrition." This identified hundreds of blogs, so the next step was to develop inclusion criteria so I could closely analyze one homogenous community. First, the bloggers' profiles in the blogs identified were required to include a dedication to promoting healthy eating. I immediately noticed how many of these blogs used the description "healthy living blog," and thus drew upon their terminology for how I referred to the blogging community. As my study focus was on the community aspect of these blogs, I searched for blogs that were linked to one another through their blogroll, links in-text links to other blogs, or links that commented on other blogs (see Ali-Hasan & Adamic, 2007). To ensure each blogger contributed enough data, blogs were required to have been maintained for at least two months and be currently active (defined as being updated at least one week prior to data collection). Two months was selected as a suitable time period for a number of critical reasons: manageability, the exceeding high number of entries, and that this time period exceeds the average of one-to-two weeks in research employing other dietary research tools such as food diaries (Fowles et al., 2007). Further, the bloggers had to be focused on sharing their daily food intake and not simply posting recipes or sharing tips for healthy eating. Finally, the bloggers could not be discussing an eating disorder, as the intent of my study was to examine healthy dietary blogs.

These selection criteria identified a community of blogs that was chosen for further study. I did not have a specific sample size but relied upon data saturation, which involved adding new blogs to the data set until no new insights were being revealed through additional blogs (Faulkner & Trotter, 2017). Ultimately, I considered saturation was reached with a sample size of 45 blogs, all written by young adult women and linked to one another as a food blogging community. I examined all food bloggers' profiles to determine their ages, genders, and locations (though location was limited to country) and created a spreadsheet of these data. Coding of the healthy living food blogs was conducted using an inductive approach of constant comparison to examine the blogs for repeated patterns. All blogs were examined from two months prior to the start of analysis and read repeatedly for entries involving the bloggers interacting and for text describing dietary behaviors or attitudes. I analysed these entries line-by-line and coded under either *anticipated themes*, which involved aspects such as how bloggers linked to each other's blogs, or *emergent themes*, which involved such aspects as how bloggers described their own food portion sizes compared to their boyfriends/husbands. These initially developed codes were applied to later data, but some were also revised over time and new codes developed as new themes emerged. These themes were then interpreted in relation to previous literature. Following this analysis, my findings questioned if these blogs were as encouraging of healthy dietary practices as the bloggers claimed. Of particular interest was how dietary restraint behaviors and attitudes were perpetuated by participating in this healthy living food blogging community.

Researching weight-loss blogs

In our research (Leggatt-Cook & Chamberlain, 2012), we were intrigued as to how people who sought to lose weight represented themselves in blogging about their experiences. We initially explored this by reading a number of these blogs, which were easy to find on the Internet. This convinced us there was something interesting to be explored in these blogs, particularly around examining why these people blogged, and how they represented themselves through their blog posts. We then undertook a scoping analysis, and quickly discovered the extremely large numbers of weight-loss blogs online, and the need to find criteria for selecting the most suitable blogs for our purposes. Our scoping review revealed that weight-loss blogs were predominantly authored by women, perhaps because weight loss is highly salient for women, and we decided to focus our attention on women's weight-loss blogs. We then searched several sites offering relevant blog lists, such as "the top 100 weight-loss blogs," relevant blog sites, such as BlogHer.com, and blogrolls from popular weight-loss blogs. From this, we identified and reviewed about 180 weight-loss blogs and developed criteria for selecting out specific blogs for analysis. These criteria were informed by a careful reading of ethics concerning the use of identifiable material from the Internet. The criteria we developed were that the blog was publicly accessible without subscription or login, the writer identified as a woman, the blog contained some explicit statement that the writer was blogging to support their weight-loss efforts, and the blog contained substantive content (active for several years and updated regularly). This reduced the pool to 31 blogs. We ascertained that these bloggers reported a wide diversity of approaches to weight loss, and we further selected blogs to represent this diversity. We also sought to select blogs that contained reflections on blogging and its role in the author's weight-loss experiences. Using these criteria, our final list was narrowed to 10 blogs, which made a manageable data corpus for analysis.

Our analysis was informed by our detailed searching, browsing, and reading across that wide range of blogs, which provided us with a good general understanding of the weight-loss blogosphere. For analysis, we printed over 1,000 pages of relevant blog content, immersed ourselves in this material and independently made notes, then shared and discussed our emerging analytical and conceptual ideas. We did not formally code the material but made copious notes on the texts, identifying ideas, concepts, meanings, and thematic issues. This process was also informed by our reading of relevant critical literature on blogging, weight loss, and subjectivity. This was followed by further joint, in-depth reading and engagement with the blogs, and from this immersion and our continuing debates and discussions, we identified, abductively, several key findings around the use of blogging to achieve weight loss.

Lessons learned

This section discusses the potentials of conducting food blog research using netnography, as well as the challenges that can be encountered in doing so.

Potentials of netnographic food blog research

An obvious potential of netnographic blog research is that we can obtain data passively and unobtrusively from blog posts and discussions that have not been influenced by the researcher's presence. Analysis of this material allows for insights into people's more naturally occurring discussions about food and eating, although we always need to keep in mind that blog content is always curated to some degree by the blogger (Davis, 2017). Further, netnographic research is both time and cost effective. Without the need to audiorecord and transcribe interviews or focus group discussions, one can obtain extensive amounts of conversational data from food blogs, quickly and in textual forms for immediate analysis. Also, the archival nature of blog entries means that they are generally automatically archived and stored, thus allowing for data collection from not only current entries and discussions from food blogs, but also to collect data from past periods, even over years. This provided a means for answering longitudinal research questions, examining the development of blogs and commentaries over time, and increasing the scope of our research (Kozinets, 2020).

Challenges of netnographic food blog research

Research on food blogs and food blogging communities raises several challenges for researchers. One important one relates to ethical considerations. Although there are several useful references on the ethics of social media research, including blogs, there is no clear consensus on the ethics of such research. Thus, it is important for researchers to spend time considering ethical issues before they begin researching food blogs. We discuss ethical considerations in greater depth in the following section. Food blog research, as with other types of social media research, is limited to examining the perspectives of people who have Internet access and are literate. Barriers, such as Internet availability and accessibility, cost, and computer literacy, do exclude certain people from this form of research (Wilkinson & Thelwall, 2011). However, at least in developed countries, the widespread availability and accessibility of social media has greatly reduced barriers to participation (Norman, 2012) and has also allowed for greater involvement of people with disabilities, mobility issues, and care-giving responsibilities (Seeman, 2008). Further challenge relates to the potential anonymity of bloggers. Food bloggers and their communities of shared interest are composed of people who might never meet outside that virtual community, people who are largely unaccountable for the information they share, and who can remain anonymous. However, netnographers argue that participants posting false information happens much less than might be expected (Kozinets, 2020; McDermott et al., 2013). Furthermore, participant anonymity can also be viewed as a positive, as it may permit participants the freedom and security to express aspects of themselves that otherwise they would not share (Kozinets, 2020).

How to conduct a food blog study

The final section of this chapter describes the steps involved in conducting a netnography of food blogs, which we discuss under three main topics: research focus, ethical considerations, and data collection and analysis.

Determine your research focus

Food blogs present intriguing data sets for researchers seeking alternative ways to learn about food and diet-related topics, although they may not always be the most appropriate or advantageous site for your research. Although food blog research can be enticing because it is less time-consuming, resource-intensive, and expensive than other, more traditional research methods, we caution researchers to not leap into food blog research without first carefully considering whether their research objectives can be answered through an examination of food blogs.

Hookway and Snee (2017) describe two broad types of research questions where blog data are appropriate, and these are also relevant for food research. The first type involves studies that focus on analyzing and interpreting the content of the food blogs. The second type uses food blogs as a way to gather data beyond the blog itself, by approaching bloggers to participate in interviews or surveys. In this chapter, we focus exclusively on the first type of question, but we encourage readers to explore the second type of question if that is relevant for them (see Hookway & Snee, 2017).

After determining the type of research question you seek to answer, you next need to determine if your topic has a sufficient online presence to provide relevant high-quality data for analysis. There need to be relevant blogs associated with your topic, these must provide sufficient data for your purposes, they must meet your data selection criteria (discussed below), and they must have appropriate access (discussed under ethics below). Also, blogs will frequently contain several different types of data beyond written text, such as images and videos. You need to decide what type of data you will collect and why it is relevant for your research aims.

Ensure your practice is ethical

Food blog research (and social media research, in general) raises questions about ethical practice. We, and many others, advocate for researchers to take a process approach to ethics, meaning that researchers should reflect on, address, and resolve ethical issues arising at every step of their research (Markham & Buchanan, 2012). Gaining approval from an ethics committee or ethics review board is only one step in ethical practice, and researchers need to be aware of ethical considerations throughout their research.

There are several articles discussing ethics in social media research that can assist researchers conducting netnographic research (British Psychological Society, 2017; Kozinets, 2020; McDermott et al., 2013; Markham & Buchanan, 2012; Paechter, 2012; Tiidenberg, 2018). In particular, Markham and Buchanan (2012) present a number of questions for researchers to consider before conducting research into blogs (see ethics text box for full list of questions). An important consideration is whether the blog would be presumed by the writer to be a safe space for sharing and support and is restricted to a small community (Markham & Buchanan, 2012). While it might appear that most blogs are public and seek an audience, this is not always the case, and determining the boundary between the public and the private is not always an easy decision, especially when the blogs of interest are about sensitive topics (Tiidenberg, 2018). In managing the considerations of public versus private, Markham and Buchanan (2015) propose that researchers need to carefully consider participants' expectations, reflecting on the sensitivity and potential vulnerability of both people and information (data), and always with a primary focus on doing no harm.

Another ethical consideration relates to the quoting of material from blog entries and comment sections. One recommended strategy to protect the people who posted information online is to provide few details on the participants and not reveal the names of the social media sites where data were collected (Wilkinson & Thelwall, 2011). However, quotations from social media could still be typed into search engines and lead the reader directly to the original discussion, "revealing" the participant. As an additional caution, researchers should examine participants' online profiles (if available) to determine whether they are providing personally identifying information in their profiles.

To include only food blogs deemed as public, we recommend the following inclusion criteria: include blogs only if they do not to require membership, registration, sign-in, or a password, and are publicly accessible through an Internet search engine. Additionally, while blog data will inevitably involve bloggers' personal opinions, it is unlikely that food blogs contain controversial or sensitive topics or opinions that would result in harm for participants (McDermott et al., 2013; Paechter, 2012; Wilkinson & Thelwall, 2011), so these may be easier to work with than blogs on more sensitive topics, such as recreational drug use or self-harm.

Collect and analyze your data

Although obtaining data from blog sites is rapid and relatively easy, you should be aware that filtering, organizing, and storing the vast amount of data online can be time-consuming (Wilkinson & Thelwall, 2011). The most common blog platforms where researchers can find blog posts that are typically longer and therefore well suited for qualitative analysis are WordPress, Blogger, and LiveJournal. To locate food blogs on these platforms, you can use the search engine within the blog or search through the interests listed on the blogger's profile page. For example, using LiveJournal's search engine and typing in the phrase "intermittent fasting" will search for blogs containing that phrase. Also, social media data have a fluid character; blogs and blogposts can be deleted or altered at any time (Weller et al., 2018). It is therefore important to ensure you have archived your data for analysis as you collect it form the websites.

After locating food blogs of interest, the next step is to develop study inclusion criteria that pertain to your research questions and aims for the study. These will be specific to your project, but we have earlier given examples from our own research on healthy living and weight-loss blogs that may be helpful to review and consider (see also, Weller et al., 2018).

Once you have identified a sample of blogs, you need to ensure that you have high-quality data. We recommend that you follow Wallendorf and Belk's (1989) well-developed field research techniques, including prolonged engagement, persistent observation, and researcher introspection. It is essential to spend an extended amount of time reading the food blogs. Other recommended practices for ensuring credible data involve selecting blogs that have user profiles, examining the content of individual posts to check for consistency over time, and examining a number of blogs to review the scope of consistency and variability.

In terms of analyzing the collected food blogs, conventional qualitative methods of text-analysis like narrative analysis, discourse analysis, and thematic analysis are all suitable, depending on the aims of your research. If your research includes images, then you will need to consider how best to analyze those (see Banks, 2018; Mannay et al., 2019). Analysis will follow the usual steps of interpretative qualitative research, starting as soon as you commence reading and reviewing blogs for data collection, and proceeding through writing notes in your research journal, becoming thoroughly familiar with your data set by reading, coding and note-taking, determining the major thematic ideas present in the data, and formulating an interpretation for presentation as your findings, offering new insights into the food-blogging phenomenon that you have chosen to investigate (see Marotzki et al., 2014).

Conclusion

The purpose of this chapter was to introduce food researchers to food blog research using netnography. We began by discussing food blogs and their distinctive characteristics, such as flexibility and communal aspects, and how they encourage users to produce material and participate in discussions in ways that traditional research methods cannot. We then outlined netnography, considered its strengths and limitations, and provided summaries of our own food blog research. Then, we discussed the potentials and challenges of conducting research with food blogs. Finally, we discussed the key steps involved in conducting a food blog study. Food blog research offers a significant and complementary addition to other research approaches currently used for digital food study researchers.

Dos and don'ts

Do determine a clear research focus for your project, and keep checking to ensure you stay with it

Do spend an extended amount of time in the initial phases of data collection and analysis by searching for food blogs, and reading and re-reading food blogs

Do ensure you have sufficient, relevant content online to fulfil your research focus and aims (remembering that a case study of a single blog may be adequate)

Do control the size of your data corpus, and don't allow it to get too large

Do keep data collection and analysis focused around your aims

Don't assume you have more to analyze than the text (unless you also interview bloggers)

Do examine food bloggers' profiles and entries for basic demographic data, such as gender, age, and location

Do review the ethics of your research, and especially consider the public/private boundary issues carefully

Do treat food blog research like any other type of qualitative research and keep a detailed research journal with an account of your decisions

Now you try

Do a keyword search for a food blogging community on a blog platform, for example, "paleo" or "vegetarian." Once you have identified a few different blogs, attempt an analysis of a few entries, thinking about what language is used, the ideas that dominate the discussion, and the opinions that bloggers are trying to convey. Can you see any patterns?

Questions for reflection

Have you ever experienced challenges with recruiting research participants through more "traditional" methods? How can blog entries complement other data collection?

Can you think of how collecting food blog data may potentially harm participants? In answering this question, consider: Are you researching a particular diet topic that would be considered sensitive? Are the participants vulnerable?

Why do you think a food blogger might reveal more online than offline?

Ethical considerations

Could analysis, publication, redistribution, or dissemination of content harm the subject in any way?

If the content of a subject's communication were to become known beyond the confines of the venue being studied would harm likely result?

Does the author/participant consider personal network of connections sensitive information?

Does author/participant consider the presentation of information or venue to be private or public

Do the terms of service conflict with ethical principles?

Is the author/subject a minor?

(Questions directly from Markham & Buchanan, 2012)

Further readings

Association of Internet Researchers. Available at http://aoir.org/

Hookway, N., & Snee, H. (2017). The blogosphere. In N. Fielding, R. Lee, & G. Blank (Eds.), *The Sage Handbook of Online Research Methods* (pp. 380–398). London: Sage.

Kozinets, R. (2020). *Netnography: The Essential Guide to Qualitative Social Media Research* (3rded.). London: Sage.

Tiidenberg, K. (2018). Ethics in digital research. In U. Flick (Ed.), *The Sage Handbook of Qualitative Data Collection* (pp. 466–479). London: Sage.

Weller, W., Bassalo, L., & Pfaff, N. (2018). Collecting data for analyzing blogs. In U. Flick (Ed.), *Sage Handbook of Qualitative Data Collection* (pp. 482–495). London: Sage.

References

Ali-Hasan, N., & Adamic, L. (2007). *Expressing social relationships on the blog through links and comments*. Proceedings of the International Conference on Weblogs and Social Media (*ICWSM*), Boulder, CO.

Baker, J., & Moore, S. (2008). Blogging as a social tool: A psychosocial examination of the effects of blogging. *Cyberpsychology and Behavior*, 11(6): 747–749.

Banks, M. (2018). Using visual data in qualitative research. In U. Flick (Ed.), *Sage Qualitative Research Kit* (Vol. 5, 2nd ed.). London: Sage.

British Psychological Society. (2017). *Ethics Guidelines for Internet-mediated Research*. INF206/04.2017. Leicester: British Psychological Society. Retrieved from https://www.bps.org.uk/sites/bps.org.uk/files/Policy/Policy%20-%20Files/Ethics%20Guidelines%20for%20Internet-mediated%20Research%20%282017%29.pdf

Davis, J. (2017). Curation: A theoretical treatment. *Information, Communication & Society*, 20(5): 770–783.

de Laat, P. (2008). Online diaries: Reflections on trust, privacy, and exhibitionism. *Ethics and Information Technology*, 10(1): 57–69.

Elliott, H., Squire, C., & O'Connell, R. (2017). Narratives of normativity and permissible transgression: Mothers' blogs about mothering, family and food in resource-constrained times. *Forum Qualitative Research*, 18(1). Retrieved from http://www.qualitative-research.net/index.php/fqs/article/view/2775

Fairclough, N. (2003). *Analysing Discourse: Textual Analysis for Social Research*. London: Routledge.

Faulkner, S. L., & Trotter, S. P. (2017). Data saturation. In J. Matthes, C. S. Davis, & R. F. Potter (Eds.), *The International Encyclopedia of Communication Research Methods* (pp. 1–2). Hoboken, New Jersey: Wiley & Sons.

Fowles, E., Sterling, B., & Walker, L. (2007). Measuring dietary intake in nursing research. *Canadian Journal of Nursing Research*, 39(2): 146–165.

Gunter, B. (2009). Blogging – Private becomes public and public becomes personalized. *Aslib Proceedings: New Information Perspectives*, 61(2): 120–126.

Gurak, L., & Antonijevic, S. (2008). The psychology of blogging: You, me, and everyone in between. *American Behavioral Scientist*, 52(1): 60–68.

Hart, D. (2018). Faux-meat and masculinity: The gendering of food on three vegan blogs. *Canadian Food Studies/La Revue canadienne des études sur l'alimentation*, 5(1): 133–155.

Hookway, N., & Snee, H. (2017). The blogosphere. In N. Fielding, R. Lee, & G. Blank, (Eds.), *The Sage Handbook of Online Research Methods* (pp. 380–398). London: Sage.

Howarth, A. (2017). Challenging the de-politicization of food poverty: Austerity food blogs. In Y. Ibrahim (Ed.), *Politics, Protest and Empowerment in Digital Spaces* (pp. 124–142). Hershey, Pennsylvania: IGI Global.

Kozinets, R. (2017). Netnography: Radical participative understanding for a networked communications society. In C. Willig & W. Stainton Rogers (Eds.), *The Sage Handbook of Qualitative Research in Psychology* (2nd ed., pp. 374–380). London, England: Sage.

Kozinets, R. (2020). *Netnography: The Essential Guide to qualitative Social Media Research* (3rd ed.). London, England: Sage.

Leggatt-Cook, C., & Chamberlain, K. (2012). Blogging for weight loss: Personal accountability, writing selves, and the weight-loss blogosphere. *Sociology of Health & Illness*, 34(7): 963–977.

Lepkowska-White, E., & Kortright, E. (2018). The business of blogging: Effective approaches of women food bloggers. *Journal of Foodservice Business Research*, 21(3): 257–279.

Li, C. (2007). *How Consumers use Social Networks*. Cambridge, MA: Forrester Research.

Lofgren, J. (2013). Food blogging and food-related media convergence. *M/C Journal*, 16(3). doi: www.journal.mediaculture.org.au/index.php/mcjournal/article/view/638

Lynch, M. (2010). Healthy habits or damaging diets: An exploratory study of a food blogging community. *Ecology of Food and Nutrition*, 49(4): 316–335.

Lynch, M. (2012). From food to fuel: Perceptions of exercise and food in a community of food bloggers. *Health Education Journal*, 71(1): 72–79.

Mannay, D., Fink, J., & Lomax, H. (2019). Visual ethnography. In P. Atkinson, S. Delamont, A. Cernat, J. Sakshaug, & R. Williams (Eds.), *Sage Research Methods Foundations*. London, England: Sage.

Markham, A., & Buchanan, E. (2012). Ethical decision-making and internet research: Recommendations from the AoIR ethics working committee (Version 2.0). Retrieved from aoir.org/reports/ethics2.pdf

Markham, A., & Buchanan, E. (2015). Ethical considerations in digital research contexts. In J. Wright (Ed.), *Encyclopedia for Social & Behavioral Sciences* (pp. 606–613). Waltham, Massachusetts: Elsevier.

Marotzki, W., Holze, J., & Verständig, D. (2014). Analysing virtual data. In U. Flick (Ed.), *Sage Handbook of Qualitative Data Analysis* (pp. 450–463). London: Sage.

McDermott, E., Roen, K., & Piela, A. (2013). Hard-to-reach youth online: Methodological advances in self-harm research. *Sexual Research Sociological Policy*, 10: 125–134.

Miller, C., & Sheperd, D. (2004). Blogging as social action: A genre analysis of the weblog. In L. Gurak, S. Antonijevic, L. Johnson, C. Ratliff, & J. Reyman, (Eds.), *Into the Blogosphere. Rhetoric, Community, and Culture of Weblogs*. Retrieved from blogs.ubc.ca/ewayne/files/2010/03/A-Genre-Analysis-of-the-Weblog.pdf

Morrison, A. (2011). Suffused by feeling and affect: The intimate public of personal mommy blogging. *Biography*, 34(1): 37–55.

Norén, L. (2011). Food blog study descriptive statistics, Part 1. *Graphic Sociology*. Retrieved from https://thesocietypages.org/graphicsociology/2011/09/29/food-blog-study-descriptive-statistics-part-1/

Norman, C. (2012). Social media and health promotion. *Global Health Promotion*, 19(3): 3–6.

Paechter, C. (2012). Researching sensitive issues online: Implications of a hybrid insider/outsider position in a retrospective ethnographic study. *Qualitative Research*, 13(1): 71–86.

Pink, S. (2016). Digital ethnography. In S. Kubitschko & A. Kaun (Eds.), *Innovative Methods in Media and Communication Research* (pp. 161–165). Cham: Palgrave Macmillan.

Potter, J. (2002). Two kinds of natural. *Discourse Studies*, 4: 539–542.

Rodney, A. (2018). Pathogenic or health-promoting? How food is framed in healthy living media for women. *Social Science & Medicine*, 213: 37–44.

Salvio, P. (2012). Dishing it out: Food blogs and post-feminist domesticity. *Gastronomica*, 12(3): 31–39.

Sanford, A. (2010) "I can air my feelings instead of eating them": Blogging as social support for the morbidly obese. *Communication Studies*, 61(5): 567–584.

Schradie, J. (2012). The trend of class, race, and ethnicity in social media inequality. *Information, Communication and Society*, 15(4): 555–571.

Seeman, N. (2008). Web 2.0 and chronic illness: New horizons, new opportunities. *Healthcare Quarterly*, 11(1): 104–110.

Strong, K., Parks, S., Anderson, E., Winett, R., & Davy, B. (2008). Weight gain prevention: Identifying theory-based targets for health behavior change in young adults. *Journal of American Dietetic Association*, 108(10): 1708–1715.

Taylor, N., & Keating, M. (2018). Contemporary food imagery: Food porn and other visual trends. *Communication Research and Practice*, 4(3): 307–323.

Tiidenberg, K. (2018). Ethics in digital research. In U. Flick (Ed.), *The Sage Handbook of Qualitative Data Collection* (pp. 466–479). London: Sage.

Véron, O. (2016). From Seitan Bourguignon to Tofu Blanquette: Popularizing veganism in France with food blogs. In J. Castricano & R. Simonsen (Eds.), *Critical Perspectives on Veganism* (pp. 287–305). Cham: Palgrave Macmillan.

Wallendorf, M., & Belk, R. (1989). Assessing trustworthiness in naturalistic consumer research. In E. Hirschman (Ed.), *Interpretive Consumer Research* (pp. 69–84). Provo, UT: New Jersey Association for Consumer Research.

Weller, W., Bassalo, L., & Pfaff, N. (2018). Collecting data for analysing blogs. In U. Flick (Ed.), *Sage Handbook of Qualitative Data Collection* (pp. 482–495). London: Sage.

Wilkinson, D., & Thelwall, M. (2011). Researching personal information on the public web: Methods and ethics. *Social Science Computer Review*, 29(4): 387–401.

10

HOW TO STUDY ONLINE RESTAURANT REVIEWS

Camilla Vásquez and Alice Chik

A brief history of restaurant reviews

Once a privileged domain restricted to a handful of elite tastemakers, today – because of online reviewing platforms – everyone has the opportunity to be a food critic.

When it comes to restaurant reviews, much has changed from the pre-digital era to the digital age. Historically, some of the earliest restaurant reviews were the result of a French tire company intending to get more people travelling on the road in their cars. Now considered by many to be the most prestigious restaurant reviewing system, few people outside of the restaurant industry may actually be aware of the Michelin star system's more humble origins (Lane, 2014). Eventually, over the course of the twentieth century, restaurant reviewing would evolve into a specialized genre of writing practiced by a handful of food writers employed by newspapers, travel guides, and trade publications. By the late twentieth century, the Zagat restaurant surveys offered one of the first opportunities for everyday consumers to weigh in and to add their opinions to those of restaurant experts.

However, a major shift occurred when the advent of the Internet suddenly enabled interested amateur writers and "foodies" to create food blogs, many of which featured restaurant reviews. Suddenly, these review texts had the potential to reach a wide readership of similarly interested food consumers.

Furthermore, the last two decades – the era associated with so-called Web 2.0 technologies – have witnessed an explosion in the popularity of online consumer reviewing platforms, some of which are either dedicated to (e.g., *OpenRice*, *Tablog*), or strongly associated with (e.g., *Yelp*), restaurant reviews. In recent years, online consumer reviewing platforms have completely transformed the practice of restaurant reviewing. What was once an activity carried out by a handful of elite experts is now open to anybody with an Internet connection. We view this development as democratizing the production of information (i.e., in the form of user-generated online content), and as running parallel with current trends in the democratization of food and taste cultures, more generally (Johnston & Bauman, 2010).

Like any other type of user-generated online content, online restaurant reviews are associated with several pros and cons. In terms of their positive effects, it has been demonstrated that restaurant reviews can be good for business (Luca, 2011), with both businesses and consumers benefitting from this form of "electronic word of mouth." From the consumer perspective, online restaurant reviews provide users with vast numbers of first-hand accounts from a wide range of other consumers. However, on the negative side, there are also numerous media reports of some users who "game the system" and engage in illegal activities, such as posting fake positive reviews in order to boost a restaurant's reputation, or conversely, posting fake negative reviews to make a competitor's business look bad. From the review users' perspective, beyond sifting through individual reviews and trying to guess whether any of them are cases of opinion fraud, or "astroturfing," there is also the issue of information overload for review readers who must make sense of the many available, and often conflicting, opinions about different restaurants.

Today, plenty of research has been published about online restaurant reviews. Of course, not all of these studies deal with food specifically. For instance, one of the earliest and much-cited studies of restaurant reviews (Luca, 2011) examined the economic impact of *Yelp* reviews on restaurant sales in one North American market and found that a one-star increase in *Yelp* ratings resulted in a 5–9% increase in a restaurant's revenue. In another domain of inquiry, scholars from computational fields have taken a big data approach and have often focused on the development of models or classifiers for carrying out opinion mining, or sentiment analysis, of reviews. Another productive strand of research from consumer studies has been concerned with readers' perceptions of online reviews. And still other studies have explored important topics ranging from using online restaurant reviews to predict food hygiene issues and inspections (e.g., Harrison et al., 2014; Kang et al., 2013) to those examining relationships between restaurant reviews, race, and neighborhood gentrification (Zukin, Lindeman, & Hurson, 2017).

Empirical work about food in restaurant reviews

Where food in online restaurant reviews is concerned, empirical research has addressed online restaurant reviews from a variety of perspectives. For instance, one study which employed the methodology of content analysis and examined over 2,000 online comments about London restaurants found that, of all the aspects of the dining experience (e.g., service, ambiance, price), food was – perhaps unsurprisingly – the most salient in users' comments (Pantelides, 2010), closely followed by service. This study also found that, of the entire meal, the starter tends to leave a lasting impression on consumers. In another recent study, which took a big data approach, researchers also found that in online reviews, food and service affect the numeric star ratings given by reviews more than price and ambiance do (Gan, Ferns, Yu, & Jin, 2017).

As scholars of language, our own interest has to do with studying the actual *language* of restaurant reviews, especially with how reviewers use language to describe food. There are currently only a handful of published studies which have addressed this topic. In one of these studies, which took a computational linguistic approach, Jurafsky et al. (2014) analyzed over 800,000 *Yelp* restaurant reviews. Their study found that many consumers framed their comments about fried, starchy, or sweet fatty foods in terms of addiction metaphors (e.g., *these cupcakes are like* crack, *the wings are* addicting) – and that this was especially the case for less expensive restaurants in their sample. In contrast, reviewers writing about food served in more expensive restaurants tended instead to rely more on metaphors of sex (e.g., sexy *food*, orgasmic *dessert*). In addition, the researchers found that such class-based differences are also simultaneously realized at the level of linguistic expression: reviewers of more expensive restaurants wrote longer reviews and used more complex vocabulary than reviewers of less expensive restaurants.

Our own work has also focused on the language of restaurant reviews (e.g., Chik & Vásquez, 2017; Vásquez & Chik, 2015; Vásquez, 2014a; Vásquez, forthcoming). For instance, in Vásquez and Chik (2015), we considered how restaurant reviewers use language to position themselves as authorities on culinary matters – as well as how they construct different kinds of "foodie" identities – in their reviews of Michelin-rated restaurants located in New York and Hong Kong. In Chik and Vásquez (2017), we provide a more comprehensive comparative analysis of reviews from online reviewing sites associated with the two locations, *Yelp* and *Open Rice*. In this study, we found that Hong Kong's *Open Rice* restaurant reviewers were much more likely to mention specific dishes in their reviews and that they also tended to be more detailed in describing food taste, smell, and appearance than their New York counterparts. In contrast, many English language *Yelp* restaurant reviewers provided only general overall evaluations of the food as (i.e., as *good, ok, not bad*), instead of going into details about its taste, texture, appearance, and so on.

We have worked in this field by applying the tools of discourse analysis to online restaurant reviews. In other words, we have examined not only the contents of review texts, but we have also focused our analytic attention on what particular forms of language reviewers have chosen to use. What this means is that we pay attention not only to *what* reviewers are saying about food, but we also pay very close attention to *how* they say it. Discourse analysis is the methodological approach that we use to do this. At the end of this chapter, interested readers can find references for several introductory textbooks that provide helpful guidelines for "how to do" discourse analysis (e.g., Gee, 2014; Johnstone, 2018; Paltridge, 2012) in the "Suggestions for Further Reading" box.

In the next section, we provide a discussion of two case study examples of our own restaurant review research, and we provide extra information detailing some of our methodological decision-making. The first of these studies addressed how reviewers describe negative food experiences and how businesses respond to those claims (Vásquez, forthcoming). The second study compared consumers' food descriptions from two different reviewing platforms associated with two different geographical contexts (Chik & Vásquez, 2017). Based on our own experiences with doing this type of research, we offer some practical recommendations related to research design, data collection, data management, and data analysis.

Example 10.1

In a recent study by Vásquez (forthcoming), one of the authors wanted to find out more about "food debates" taking place on online restaurant reviewing platforms. In other words, what kinds of language do customers use to discuss food-related problems – and how do the affected businesses respond to those food-related complaints on those same platforms?

The first step in the research process involved narrowing down the sample. Which restaurants should be included, and how many? Fortunately, one of the local newspapers in my area had just published its annual listing of the "Top 50 Restaurants" in our region. Because I wanted to collect only those reviews that had received responses from businesses, this list seemed like a good place to start. I speculated that those establishments which had been designated as "top restaurants" would be invested in their success and were therefore likely to be active on reviewing platforms as a way of showing their interest in customer satisfaction.

Deciding which reviewing platforms to collect data from was the next decision I faced. At the time, the two most popular restaurant reviewing platforms were *Yelp* and *TripAdvisor*, so I decided to use those. Because the focus of my study was on "food disputes" – and this focus entailed discussions of unsuccessful or problematic food experiences – my next step was to identify all of the available negative (i.e., one-star or two-star) reviews for the 50 restaurants in the sample on the two platforms. There were 1,539 in total. However, the aim of the study was to compare how reviewers talked about food-related problems with how businesses addressed those same food-related problems, so the final data set needed to include only those reviews which featured a follow-up response from the business. As it turned out, this was less than 10% of the total 1,539 negative restaurant reviews, yielding a final data set of 96 pairs of reviews and related business responses (i.e., 192 unique texts). From my past experiences conducting qualitative discourse analytic research, I knew that 192 texts would likely be an appropriately sized data set with which to investigate this topic.

Data sampling and selection is one of the most important and challenging aspects of methodological decision-making facing any researcher who interested in exploring online reviews. Today, there are literally billions of reviews which have been posted online about any given topic and category, making it is impossible to study *all* of them. Consequently, it becomes necessary to impose some kind of parameters on which reviews to collect. Guided by the research questions and goals of the study, researchers of restaurant reviews will need to make principled decisions about which data they will collect and why. Being forthcoming and transparent about these important methodological decisions will help contribute to a study's reliability and its trustworthiness.

Turning once again to the practical matters of data collection and data management, in order to create the actual data set to work from, I manually downloaded these 192 texts (i.e., the total of 96 negative reviews, along with the associated responses from businesses) from *Yelp* and *TripAdvisor* and saved each of them as a Word document. Next, I devised a system of ID labels for each piece of data. I organized the data alphabetically by name of restaurant. The first restaurant in the set began with the letter "B." This restaurant had 11 negative reviews which received responses. I labelled the first review BRev_1 and its corresponding response, BResp_1. Each subsequent pair of texts was labelled BRev_2, BResp_2, BRev_3, BResp_3, and so on. This system allowed me to match up reviews and responses, as well as to cross-reference the data, and to sort the texts in different ways.

Now that the study's research questions were established, the sampling parameters were defined, and the data were collected, I was finally ready to begin my analysis. I started by identifying the descriptions of food (e.g., *The food was very poor.* or *The caesar was bland, bland, bland.* or *The white bean soup was weak and so full of pepper it was horrible.*) in each review. This step was important because the dining experience is multi-faceted, and therefore, restaurant reviews often contain information about aspects that are unrelated to food, such as location, service, ambiance, weather, price, parking, personal information about the reviewer, and so on. This step involved identifying, extracting, and eventually coding, any and all mentions of food.

For instance, Excerpt 10.1 shows a two-star review from *Yelp*. In it, the reviewer begins by providing some context for his visit and then describes the restaurant's physical space. This is then followed by complaints about seven different food items (i.e., *brisket, beans, slaw, ribs, chicken, pulled pork, apple crisp*). These complaints vary in the intensity of their assessment: for example, describing one dish as *just ok* versus another as *very disappointing*. I have highlighted below all of the food-related complaint language in Excerpt 10.1.

EXCERPT 10.1 [UREV_7]

I read all of the good reviews and so we stopped by on a Saturday morning. First of all, I must say that we came here for the BBQ. It's a small quaint place (inside) that reminded me of some bistros in Austin, Tx. There are no full sized tables inside designed to hold large trays of food but there are picnic tables in the back patio in full sun or metal tables in front. We opted to eat at the bar.

Seeing the pictures online, we were stunned as to what our BBQ looked like out of the kitchen. The brisket was nothing less than shredded pot roast, the beans and slaw were not authentic. The ribs were just OK. The chicken and pulled pork were so-so but the collards were awesome. We ordered the Apple Crisp as our dessert to share. It was very disappointing...raw apples that were cold on one side of the pan and hot on the other. In all, a poorly made dessert. Our bad luck, I guess.

Once I identified all of the food-related complaints in the restaurant reviews, I then examined the businesses responses and highlighted any references that businesses made to the reviewers' food-related complaints. Similar to the reviews, business responses often contained lots of additional information that was unrelated to food, so it was important to pinpoint only the part of the text that addressed food issues. Excerpt 10.2 below shows entire the response that was posted on *Yelp* by the Business Manager of the complained-about restaurant in Excerpt 10.1. I have highlighted the part of the response that corresponds to the reviewer's food-related complaint language. As you can see, this business representative chose to address only one of the seven complained-about food items in the customer review (i.e., *brisket*) in his explanation for why that particular item did not meet the reviewer's expectations.

EXCERPT 10.2 [URESP_7]

Hey [*name*], sorry to hear you didn't have a good experience. <u>It sounds like your piece of brisket was off the point instead of the flat. The point is higher in fat content and doesn't slice neatly so it can resemble a roast or even shoulder meat.</u> Thank you for your feedback and hopefully you have a better experience next time you visit.

Once I completed this text-extraction process for the entire data set, I copied and pasted all of the highlighted texts into an Excel file. Doing this enabled me to gain a big-picture overview for food-related language in the data, and it also facilitated data management and manipulation. Because an ID label was included with each text fragment, this allowed me to cross-reference back to the original Word files, in case I needed more context.

My data set was relatively small (96 reviews + 96 responses = approximately 120 pages of text), so I conducted all of the analytic procedures described above manually. However, researchers working with larger data sets might find it helpful to work with a CAQDAS (Computer-Assisted Qualitative Data Analysis Software) package in order to extract, code, label, and sort their data. Some of the most popular software programs for qualitative analysis include MAXQDA, NVivo, and Atlas-TI. Paulus, Lester and Dempster (2014) offer helpful guidelines for selecting a CAQDAS package.

Speaking of the size of a data set, one issue facing researchers carrying out an analysis of review texts is the question of "How much data is enough?" In the case of this study, I was able to detect several meaningful patterns with the 96 review/response pairs, and I felt confident that this gave me a reasonably representative sample of what goes on in "food debates" between reviewers and businesses in online reviewing platforms. The exact amount of data necessary to carry out an analysis, of course, will depend on the research questions guiding the study. Many qualitative researchers use to the notion of "data saturation" (Saunders et al., 2018) to describe the point at which no new information or themes are observed in the data. Reaching data saturation is an indication that enough data have been collected and analysed to meaningfully address the study's research questions.

After completing my analysis, I discovered some interesting patterns. Nearly half of reviewers described their food in only very general terms (e.g., *my meal was awful*; *the food was mediocre*). Yet – as we saw above in Example 10.1 – some other reviewers provided more specific details about one or more dishes. However, within these descriptions of specific dishes, I discovered that there was actually a relatively limited set of adjectives used by reviewers to describe unsatisfying or unpleasant food experiences. Additionally, I found that a handful of these (such as *bland*, *flavorless*, *salty*, *weak*, *dry*, *greasy*, *soggy*, and *raw*) appeared repeatedly in several different reviews throughout the data set.

As mentioned, I found that descriptions of food in reviewers' complaints varied between being general or referring to specific dishes. In contrast, I found that restaurant responses patterned quite differently. Most often, in their responses, restaurant staff either avoided any mention at all of food, preferring instead to use the general noun "experience" to encompass the restaurant visit more broadly (as in, for instance, a response such as: *I'm sorry to hear about your unfortunate experience recently at* [restaurant name].) This makes sense, because many of the reviewers who complained about food also tended to complain about other dimensions of their restaurant experience (e.g., service, ambiance, wait time). And in the fewer cases in which restaurants did take up food more specifically, the dominant tendency was to keep the wording fairly vague (e.g., *We're sorry that your meals and service were less than stellar.*). In this sense, Excerpt 10.2 above differs from the majority of restaurant responses, in that the business representative opted to address at least one specific food-related complaint in his text.

Considering that reviewers are providing food-related accounts that are based on their own personal experiences, they have the ability to provide detailed and specific evaluations that are based on their first-hand impressions, memories, and recollections. In contrast, restaurant responses are written by a wide range of personnel occupying many different professional roles (e.g., owners, general managers, social media managers – and even second parties such as employees from communication firms who are contracted to handle social media and public relations on behalf of the restaurants). These individuals who represent the business may or may not have even been present during the reviewer's visit. For this reason, they may not always be able to address a food-related issue based on their personal experience, which may explain why their responses tend to be vague or general. Moreover, in some cases, they may deliberately choose to avoid repeating a food-related issue in their response, as a kind of "impression management" strategy. Obviously, it is never in the restaurant's best interest to agree that the quality of their food was somehow not good; however, disagreeing politely with the customer's complaint in a public online forum can sometimes prove to be quite challenging.

Example 10.2

In our article, "A comparative multimodal discourse analysis of restaurant reviews from two geographical contexts," we were interested to find out more about how restaurant reviewers writing from two different cultural contexts (New York and Hong Kong) evaluated their negative experiences of dining in Michelin-star restaurants. Though reviewers came from two major cities, the reviews revealed similar dining expectations for this category of restaurant. Yet, we found distinctively different discursive practices. First, Hong Kong reviewers clearly prioritized the quality of food over quality of service. And this was evidenced in their detailed descriptions of the dishes, the cooking method, and their sensory experiences (Excerpt 10.3). (In order to be able to compare the New York reviews with the Hong Kong reviews, one of the authors had to first translate the Hong Kong reviews from Chinese into English. We discuss this process – as well provide explanations for why Hong Kong reviewers were more descriptive about food – in more detail in Chik and Vásquez (2017).

EXCERPT 10.3

The quality of the crispy tofu: 12 pieces of fried tofu were like gold bricks, neatly placed on a rectangular plate, with a few leaves of decoration. I'd call it ordinary but nice. The surface of the tofu was fried to crispness, it wasn't greasy at all, and not too salty, just right. But it's a pity that local frozen tofu was used. Although it was very smooth, there was no soya taste. [OpenRice]

Second, the Hong Kong reviewers were also much more likely to include multimodal components such as emojis and photographs in their reviews (Excerpt 10.4).

EXCERPT 10.4

Tofu with crabmeat – (VERY VERY BAD), the tofu was tasteless, felt like it was just put there, not cooked with the crabmeat  and the orange colour was a bit strange . [OpenRice]

In addition, we also found that when Hong Kong reviewers chose to include photos in their reviews, the photos were rarely referenced in the review texts. Many of these photos were assumed to be self-explanatory references of the dishes mentioned in the written reviews.

Potentials and challenges

As we mentioned earlier, the millions of user-generated restaurant reviews available online provide digital food scholars many potentials as well as many challenges. As far as potentials are concerned, the studies we have discussed so far in this chapter here have only scratched the surface of food in restaurant reviews. There is much more research to be done in this area, and below we point out several potential topics for future research.

In terms of challenges, as we have already noted, it can be quite difficult to restrict the scope of a study and to sample from all available review data according to a set of principled parameters. For example, if we want to learn more about a topic like "What types of food photos posted on reviewing sites are most likely to entice readers to visit a restaurant?" we would need to begin by figuring out how to best narrow our scope to the most relevant reviewing platform, the most relevant set of reviews, and the most relevant review photos. Where and how should we begin to address this study? Where and what kind of data would be best to collect? And how much data will we need to address our question? In addition, to address a complex research question like this one, we would also need to combine some analysis of images (see Ledin & Machin, 2018, for guidelines) with interviews from people who actually use restaurant reviews to guide their consumer decision-making.

Another challenge facing review researchers is the temptation to collect a massive amount of review data, precisely because it is so readily available. However, having more data does not necessarily result in a better study. Furthermore, in order to carry out any kind of meaningful analysis with big data sets (e.g., Jurafsky et al., 2014), it is necessary to have advanced computational skills to analyse those data. Furthermore, collecting big data sets also requires special tools for data collection, such as web crawlers, which are programs written to automatically collect data from a given site. However, the use of web crawlers violates the legal "terms of service" of many online reviewing platforms. In contrast, collecting data manually (i.e., "copying and pasting") for research purposes is generally considered to be "fair use" and does not violate platform's terms of service.

Lessons learned

In addition to posting written reviews in which restaurant guests describe and evaluate their visit, most reviewing platforms also require reviewers to assign a numeric rating assessing their overall experience. Rating scales on reviewing platforms typically range from 1 to 5 stars. Beyond assigning a numeric score, it is clear from our research that the majority of reviewers are interested in evaluating their restaurant experience within their reviews. A minimal evaluation might consist of only a very short phrase or sentence. For instance, a cursory glance at a handful of restaurant reviews *Google* turns up many examples of very short reviews such as (*"Fabulous fun and food." "Really nice spot. Great service and management." "Great small plates place."*

"Love this place!"). Reviews such as these provide an evaluation or assessment of some particular aspect/(s) of the dining experience, yet they provide readers with no specific details, or any descriptions of food.

Unlike the reviewers who offer short, general evaluations, other reviewers choose to provide much more detailed descriptions, especially about the food they consumed. And still other review authors choose to package their review texts in the form of a personal narrative, organizing their prose according to a chronological sequence of events (Vásquez, 2015). We have found that these narrative accounts of restaurant visits vary in the amount of personal information reviewers weave into their reviews. The more information they provide, the more insights readers gain into what kinds of people they are, or what kinds of things matter to them. Of course, sometimes this may be an intentional act of self-disclosure, while other times, this personal information can be unintentionally "given off" – for instance, in a word choice that gives readers a clue about the reviewer's country of origin (Vásquez, 2014b). A restaurant review that is presented as a personal narrative can be extremely concise, as for example, in one of the shortest narrative restaurant reviews we have come across: *Ordered rice with tea smoked duck for lunch for delivery. Food was good until I found a hair in it. Enough said.* (Vásquez, 2014a). Alternatively, some restaurant reviewers craft much more elaborate narratives that include surprising storylines, plot twists, and even including entire dialogues between reviewers and restaurant staff. Crafting a review text as a narrative is one of several ways of attracting readers' attention and making one's review stand out from dozens of very similar texts.

Ways of doing

In this section, we identify some important directions for further research about food in restaurant reviews. The first of these concerns research addressing the sensory descriptions of food. As noted above, a handful of studies have addressed how reviewers from different cultural backgrounds talk about food (Chik & Vásquez, 2017); how reviewers writing about different restaurant price categories talk about food (Jurafsky et al., 2014); and how reviewers versus business owners talk about food (Vásquez, forthcoming). Building on and extending this topic of food descriptions, we envision researchers continuing to advance our knowledge of language associated with notions of food appearance, taste, texture, smell, temperature, and so on.

Another "hot topic" concerns the notion of authenticity in food. In some cases, comments on authentic or inauthentic food in restaurant reviews may be bound up with discourses that are implicitly racist (e.g., Kay, 2017; Zukin, Lindeman, & Hurson, 2017). Other researchers might be interested in learning more about how restaurants located in understudied areas or contexts are evaluated in online reviews. In addition, it may also be interesting to explore how specialty foods are described (e.g., what language do wine experts use to describe wines in online reviews?). This last point could be extended to consider how expertise or credibility is constructed using language in various food or beverage categories.

Three final areas we envision where much more work could be done include the following: 1) multimodality, 2) cross-linguistic or cross-cultural analysis, 3) cross-platform analysis. For multimodality, it would be interesting to learn more about food images that reviewers post on reviewing platforms. As described in Example 10.2 above, we were interested in how reviewers from different linguistic and cultural backgrounds approached the same activity of writing restaurant reviews. Our study yielded some very interesting results, so we would like to see more cross-linguistic or cross-cultural analyses involving other languages and cultures (Interested readers may want to take a look at the work of Cenni & Goenthals, 2017, which focuses on hotel reviews, but which offers an excellent model for this type of cross-linguistic work). Finally, both our 2015 and 2017 articles necessarily involved cross-platform analysis (i.e., reviews from *OpenRice* and *Yelp*) because they addressed restaurants in two different geographical contexts. It may also be interesting to further expand this work by comparing reviews of the same restaurant/(s) from two different platforms (e.g., *TripAdvisor* and *Google*, or *Yelp* and *Facebook*).

Dos and Don'ts

Do acquaint yourself first hand with your reviewing platform's particular (a) affordances and (b) constraints: for instance, (a) the ability for users to post photos, and (b) limits on the number of photos that an individual reviewer can post.

Do understand and accept that reviewing platforms are constantly changing and making system upgrades. Any data set will necessarily only reflect the features of the platform at a very specific point in time, and platforms may even make changes to their interfaces or the functions they offer in the midst of your research project. This is a normal part of carrying out research with online data.

Don't be afraid to explore under-explored topics!

Now you try

[EXCERPT A: NEW YORK, *YELP*]

My friends all know that I'm extremely materialistic when it comes to restaurants. What exactly does that mean when it's used to describe a gastronome? It means, I make it a mission to try all the Michelin restaurants, all positive NY Times reviews and whatever restaurants have received public accolades from popular channels of food news.

[EXCERPT B: HONG KONG, *OPENRICE*]

The best dish: Shrimp with Longjing tea leaves. The river shrimp were fresh and crunchy, crystal in appearance, and cooked without any baking soda. I'd say these could be among the most delicious stir-fried shrimp I've ever had. This dish totally showcased the freshness of ingredients.

Amateur reviewers frequently use different methods to establish their expertise and their credibility.

How do the reviewers in Excerpt A and B establish their credibility?

What kinds of language would you use to display your knowledge about restaurants and food, if you were writing a review?

Have you noticed any specific language features used by reviewers writing on other reviewing platforms?

Questions for reflection

Many of studies focus on reviews written in English. Have you noticed any linguistic and non-linguistic differences in reviews written in languages other than English?

How are restaurant reviews constructed in your own cultural context?

Are there any differences in reviews written for "everyday" restaurants versus more expensive ones?

Ethical considerations

One challenge facing researchers of restaurant reviews is grappling with ethical considerations. Reviews posted on most review platforms are public texts – that is, anyone can access them. However, it is difficult to know how the authors of those texts would feel about having their data used for research purposes. In many cases, it may be appropriate to de-identify data, either by deleting any information that would link the data to the author (e.g., name or userID), or by using pseudonyms to refer to an individual author. Depending on the number of reviews that are collected for a particular study, researchers may consider contacting authors of texts to get permission for using the content as data. However, this is usually only feasible when one is working with a very small data set. The Association of Internet Researchers (AOIR) has published guidelines for ethical decision-making when working with online data. It can be helpful to consult these before beginning a research project: https://aoir.org/reports/ethics2.pdf One principle of conducting ethical research is "doing no harm" and for this reason, we do not disclose the names of any of the complained-about restaurants discussed in the studies featured in our Examples section.

Further readings

Gee, J. P. (2014). *An Introduction to Discourse Analysis* (4th ed.). London: Routledge.
Johnstone, B. (2018). *Discourse Analysis* (3rd ed.). Oxford: Wiley, Blackwell.
Paltridge, B. (2012). *Discourse Analysis: An Introduction* (2nd ed.) London: Bloomsbury.

References

Cenni, I., & Goenthals, P. (2017). Negative hotel reviews on TripAdvisor: A cross-linguistic analysis. *Discourse, Context & Media*, 16, 22–30. doi: 10.1016/j.dcm.2017.01.004

Chik, A., & Vásquez, C. (2017). A comparative multimodal discourse analysis of restaurant reviews from two geographical contexts. *Visual Communication*, 16 (1), 3–26. doi: 10.1177/1470357216634005

Gan, Q., Ferns, B., Yu, Y., & Jin, L. (2017). A text mining and multidimensional sentiment analysis of online restaurant reviews. *Journal of Quality Assurance*, 18 (4), 465–492. doi: 10.1080/1528008X.2016.1250243

Harrison, C., Jorder, M., Stern, H., Stavinsky, F., Reddy, V., Hanson, H., Waechter, H., Lowe, L., Gravano, L., & Balter, S. (2014). Using Online Reviews by Restaurant Patrons to Identify Unreported Cases of Foodborne Illness: New York City, 2012–2013. *Morbidity and Mortality Weekly Report*, Centers for Disease Control PMID: 24848215. https://www.ncbi.nlm.nih.gov/pmc/articles/PMC4584915/

Johnston, J., & Bauman, S. (2010). *Foodies: Democracy and Distinction in the Gourmet Foodscape*. London: Routledge.

Jurafsky, D., Chahuneau, V., Routledge, B. R., & Smith, N. A. (2014). Narrative framing of consumer sentiment in online restaurant reviews. *First Monday*, 19 (4). https://firstmonday.org/ojs/index.php/fm/article/view/4944/3863

Kang, J. S., Kuznetsova, P., Luca, M., & Choi, Y. (2013). *Where NOT to eat: Improving public policy by predicting hygiene inspections using online reviews. Proceedings of the 2013 Conference on Empirical Methods in Natural Language Processing* (pp. 1443–1448). Seattle, WA: Association for Computational Linguistics.

Kay, S. (2017). *Authenticity in Online Ethnic Restaurant Reviews: Revealing Conflicted Nationalism in Multicultural Consumption*. Master's Thesis, New York University.

Lane, C. (2014). *The cultivation of taste*. Oxford: Oxford University Press.

Ledin, P., & Machin, D. (2018). *Doing Visual Analysis*. London: Sage.

Luca, M. (2011). Reviews, reputation, and revenue: The case of *Yelp*.com. Harvard Business School Working Paper 12–016. http://go.mainstreethub.com/rs/mainstreethub/images/Yelp%20Study.pdf

Pantelides, I. (2010). Electronic meal experience: A content analysis of online restaurant comments. *Cornell Hospitality Quarterly*, 51 (4), 483–491. doi: 10.1177/1938965510378574

Paulus, T., Lester, J., & Dempster, P. (2014). *Digital Tools for Qualitative Research*. Thousand Oaks, Sage.

Saunders, B., Sim, J., Kingstone, T., Baker, S., Waterfield, J., Bartlam, B., Burroughs, H., & Jinks, C. (2018). Saturation in qualitative research: Exploring its conceptualization and operation. *Quality & Quantity*, 52, 1893–1907. doi: 10.1007/s11135-017-0574-8

Vásquez, C. (2011). Complaints online: The case of TripAdvisor. *Journal of Pragmatics*, 43, 1707–1717. doi: 10.1016/j.pragma.2010.11.007

Vásquez, C. (2014a). *The Discourse of Online Consumer Reviews*. Bloomsbury: London.

Vásquez, C. (2014b). "Usually not one to complain but…": Constructing identities in online reviews. In P. Seargeant & C. Tagg (Eds.), *The Language of Social Media: Community and Identity on The Internet* (pp. 65–90). London: Palgrave Macmillan.

Vásquez, C. (2015). Right now versus back then: Recency and remoteness as discursive resources in online reviews. *Discourse, Context & Media*, 9, 5–13. doi: 10.1016/j.dcm.2015.05.010

Vásquez, C. (forthcoming). What if the customer is wrong?: Debates about food on Yelp and TripAdvisor. In C. Gordon & A. Tovares (Eds.), *Identity and Ideology in Digital Food Discourse: Social Media Interactions Across Cultural Contexts*. London: Bloomsbury.

Vásquez, C., & Chik, A. (2015). "I am not a foodie…": Culinary capital in online reviews of Michelin restaurants. *Food & Foodways*, 23 (4), 231–250. doi: 10.1080/07409710.2015.1102483

Zukin, S., Lindeman, S., & Hurson, L. (2017). The omnivore's neighborhood? Online restaurant reviews, race and gentrification. *Journal of Consumer Culture*, 17 (3), 459–479. doi: 10.1177/1469540515611203

PART 3
Users' practices

11

DIGITAL FOOD TRACKING

Combining traditional and digital ethnographic methods to identify the influence of social media sharing of health and foods upon users' everyday lives

Rachael Kent

Introduction

Digitally tracking our health and lifestyle has become a pervasive everyday practice for many of us. The emergence of the smartphone, mobile applications (apps), devices, and wearable technologies have enabled a dramatic growth in the use of self-tracking technologies, and social media platforms, presenting new opportunities to mediate health, wellness, and lifestyle practices. The global mobile health app market was valued at approximately $37billion in 2019 (statistica.com) and is expected to generate around $111.1billion by 2025 (Zion Market Research, 2018). As of the beginning of 2019, there are 3.484billion active social media users globally (wearesocial.com). Tracking and monitoring health and lifestyle, however, is nothing new. For centuries, people have reflexively monitored aspects of their personal lives, daily habits, and food consumption, often in the form of keeping diaries (Rettberg, 2014). Digital applications, however, present new interactive spaces from which to capture these everyday health-related habits, and to share such data and content on a variety of platforms for other users, networks, and communities to engage with. Performing these "health(y)" identities for the communities' gaze is frequently achieved through the careful inclusion and exclusion of certain foods, and representations of health and lifestyle (Kent, 2018, 2020a). Self-tracking, therefore, can be understood as both a qualitative and quantitative practice, with data generated being both numbers and statistics related to quantifiable food and health metrics (such as calorie counting and exercise tracking), as well as photographic representations of the body, and food consumption (such as #foodporn and fitness "selfies"), which all provide the ability to capture the multi-dimensional aspects of one's health and lifestyle into aesthetically pleasing and interactive content. The sheer abundance of data and nutritional information circulated by lay people, nutritionists, clinicians, and "influencers" on health apps (like Fitbit and MyFitnessPal) and social media (such as Facebook, Instagram, and WeChat) further complicates health (mis)information within these competitive and comparative digital food cultures, as many users become "experts" on managing and optimising "good" health (Kent, 2020a). These mediated, performed, and curated practices enable a new articulation, representation, and way of understanding our bodies, health, and identity.

The influence of engaging with these platforms to manage health is vast and pervasive; at either ends of the spectrum, these practices enable successful health and diet improvement via tracking and sharing, to issues around compulsive and addictive use, and subsequent negative effects upon diet management and user's wellbeing (Kent, 2020b). These new types of communicative environments afford researchers new interactive spheres from which to identify and analyse human decision-making processes around health and food consumption within digital food cultures, and how this influences the health-related behavioural practices and decisions they make in their everyday (offline) lives. This provides a contextual depth which would be limited and potentially unachievable without the self-representational affordances of both self-tracking devices and social media. The importance of the ongoing examination into these practices from a long-term perspective cannot and should not be ignored. As Beato (2012: n.p.) highlights, "we treat even our most mundane lunches as if they were corpses at a crime scene." In order to examine these influential practices, this chapter proposes an innovative methodological approach to explore online and offline divides and negotiations by self-tracking and social media users. This enables identification of how, through the acquisition and sharing of both qualitative and quantitative self-tracking data, the body and health becomes managed and performed, and how this process subsequently influences diet and lifestyle.

This chapter, therefore, will provide guidance on "how to" do digital food cultures research on converged self-tracking and social media platforms from the perspective of how users share digital food content, and how this influences health management in their offline everyday lives. Through empirical research, this chapter presents the utility in combining (pre-digital) ethnographic methods (semi-structured interviews and guided diaries) with digital ethnography (online data capture) for analysing self-tracking practices within digital food cultures. The under-used method of guided diaries is used to tease out the users' everyday perspectives (Kenten, 2010) in their deliberation of what data they capture, share and why, which enables attention to be usefully drawn to their private minds, as well as their online public representations and performance of health. This is examined and discussed in combination with the capture of users' online data and representations of health and lifestyle to gather text from which to draw reflections, while contextualising findings through semi-structured interviews to enable a continuity and depth to the analysis and findings. This enables attention to be drawn to the critical and long-term temporal reflection on these user practices. Broadly, this triangulation and new methodological approach provides an understanding of how users increasingly mediate their lives through both self-tracking and social media platforms, contributing important empirical research currently missing from digital food cultures literature. This chapter will first briefly discuss the existing methodologies which have been commonly utilised within ethnographies of social media and self-tracking technologies.

Previous research

Research into digital food tracking has predominantly focused upon survey and interview-based methodologies. Some studies have focused on features of food-tracking apps' accuracy, usability, and accountability (Chen et al., 2015), their contribution to eating disorders (Simpson & Mazzeo, 2017; Levinson et al., 2017), and whether they can support healthier food purchasing (Flaherty et al., 2018). Research into digital food cultures on social media has explored the affects, moralities, and binaries that underpin these practices (Kent, 2020a; Lupton, 2020), the role of social media "influencers" upon food trends and individual consumption (Rousseau, 2012), the rise of food photography (Lewis, 2018), #foodporn (Lupton, 2019), as well as "'food selfies" as a popular form of content shared on social media (Mejova et al., 2016; Middha, 2018). No research into digital food tracking and social media sharing of these practices has adopted a triangulated mixed methods approach of guided diaries, semi-structured interviews and online content to explore a long-term detailed examination into user's health and food related practices both online and offline in their everyday lives. Conceiving participants' use of self-tracking and social media platforms as interactive ethnographic sites, enables interrogation of the online representations, offline realities, and lived experiences of these individuals. However, it must be highlighted that in adopting this research approach, the ethnography does not have to attend to a detailed analysis of these platforms (Facebook and Instagram) of self-representation themselves. An understanding of how their socio-technological features and affordance influences the tools individuals employ to self-present (e.g., on Instagram the use of filters and hashtags) is useful. However, detailed analysis should focus upon user interpretation of these sites; the processes and practices related to using these technologies to represent their health identities, and how such performances, under the online communities' gaze, affect their health behaviours and practices offline. Participants' use and engage with these platforms for their own representational and surveillant needs. As Ruckenstein and Kristensen (2018: 2) articulate: "These collaborations (…) mediate and modify human presence and perception, behaviour and decision making," enabling participants to generate new ways of seeing themselves through such self-representational performances on social media. This further shapes "self-understanding and self-expression, suggesting a vision of technology that in its concrete materiality influences not only selves, bodies and socialities but also communication and learning" (Ruckenstein & Kristensen, 2018: 2). Therefore, critical analysis should explore how these technologies mediate participants' perceptions of themselves and their personal health-related actions. Once this content is shared, this analysis then examines how participants' conceptualise "health"-related experience, individually and through the gaze of others watching in their social media communities. The next section will provide an overview of my existing empirical work in the field, research design, and methodology.

Example – Existing Empirical Work

This section will introduce the existing empirical work I have done in the field by presenting the practical application of the pre-digital and digital ethnographic methods I used in my research project (Ref: LRS-15/16-2156). My research was concerned with exploring the following three research questions:

Research questions

(1) How do users of self-tracking technologies and social media self-represent their bodies and "health"?
(2) How do these practices and self-representations enable ways of experiencing, understanding, and viewing one's own body and "health," in relation to others?
(3) Does the sharing of self-tracking data, images of diet, representations of the body and "healthy practices" and behaviours lead to "healthier" lifestyles or "healthier" bodies?

Aims and objectives

To achieve this, the project undertook ethnographic research over a nine-month period with 14 participants who self-selected through a call for participants on Facebook and Instagram: seven women and seven men, between 26 and 49 years of age who regularly (daily/ weekly) share health, fitness, and food-related content on Facebook and/or Instagram. These participants ranged from the everyday layperson, those who were dieting or training for marathons, to those dealing with illness or disease. The content shared came in the form of self-tracking data from applications (e.g., Nike+ or Strava) and devices (e.g., Fitbit or Garmin Watch), gym or fitness selfies, or more general "healthy" self-representations such as food photography. The triangulated methodological approach was as follows:

1. Two semi-structured interviews (30–45 minutes each), one pre and one post the diary period.
2. Bi-monthly guided diary entries (six entries in total) on different days of the week, over a period of three months.
3. Screenshots from content shared on Facebook and/or Instagram on the day the guided diary was completed (supplied by the participant in the guided diary).
4. Textual and thematic analysis of the language used in verbal (interviews) and written (online data and guided diaries), and screenshots of visual content shared (images and photographs).

Although these types and modes of content differ in their qualitative and quantitative capture, socio-technological affordance, and representational states, they all enable participants to engage with the self-representation of their health, and food-related practices. It should be noted at this stage that this chapter will not outline a preferable analytical method (e.g., discourse, textual, or thematic analysis) from which to analyse the data from interviews, diaries, and online content as this will depend on the research questions and focus of the individual project. The following sections will outline my experiences of conducting this innovative methodological approach in digital food cultures, what I learnt, strengths and challenges, as well as guidance dos and don'ts for those wishing to conduct a similar mixed-methodological approach in this field.

How to do (digital) ethnography on social media and self-tracking for a digital food study

(Digital) ethnography

The next section considers some aspects of "digital ethnography" related to the methodologies used in this type of research approach. It is important to recognise at this stage, however, that by triangulating interviews, guided diaries, and online data, this research approach would not consider itself a solely "digital ethnography" in the sense that it does not examine the digital platforms themselves as ethnographic sites to be analysed in detail. Analysis focused upon the participants processes of food-related content sharing and how this influenced daily health, diet, and lifestyle. Examining only online data could limit context within the analysis, removing recognition of where it came from; it can only speculate as to "why" that content was produced, without uncovering the practices and processes users employ to produce such content. Therefore, this research approach overcomes distinctions between "online" and "offline" communications, not seeing them as overtly challenging or difficult to capture or analyse, but the analysis instead centres around the user's interpretations of these sites. The platforms, applications, and devices (self-tracking technologies and social media), in which the users are engaging, are of course digital; self- and peer-surveillance operates (post the imaginary perceptions of the users) through these digital spaces, and the constructed representations of health identity (screenshots of shared content) are curated and documented on them. However, rather than focusing in detail upon the digital platforms themselves, this research analyses the user's interpretations of their engagement. Therefore, in the context of triangulating these methodologies, ethnography does not examine the digital through the mass collection of data. Depending on the researcher's questions, that could be problematic, as ethnographically speaking, those data are taken out of context. Rather, the user's representations of themselves and their "health," food, and lifestyle through the online content shared is contextualised through the other methodologies of interviews and guided diaries.

Ethnography "allows us to situate consumption practices and to contextualise media use in everyday cultural life" (Mackay, 2005: 131). However, media ethnography has been criticised for undertaking "hit and run" analyses (Murphy & Kraidy,

2003: 12), contributing to "partial truths" (Moore, 1993: 4). Triangulating mixed methods over period of time (in my case nine-months) enables the researcher to cross-reference data and to ensure that analysis does not fall victim of providing a partial investigation of ethnographic accounts. Analysis of the online content (provided as a screenshot of daily/weekly sharing) enabled an examination of a carefully curated self-representational text of their bodies, health-related activities, and lifestyles. Analysis of the interviews and guided diaries enabled an examination of how participants and online communities interpreted that text through "lived experience" (McRobbie, 2009 [1992]), in their everyday lives, both online and offline. The long-term temporal nature of these triangulated methods helped to grasp cultural understandings through layers of data, to improve the analysis.

Digital ethnography, therefore, celebrates the capturing of online data as a way of providing insights into digital spheres, without always attending to the recognition that platforms relationship with users are an interactional and participatory event. Therefore, triangulating these research methods ensures distinctions between online data and representations made by participants are not drawn, or identified as separate from their offline worlds. Distinguishing between online and offline spheres when researching social media and self-tracking practices can be restrictive; "activities in these realms become increasingly merged in our society and the two spaces interact with and transform each other" (Garcia et al., 2009: 53). This research approach identifies that the online and offline modes of communication are different in regard to the methodological capture (interviews, guided diaries, and online content), but not in terms of the analytical examination of these methods. Different critical forms of analysis can be utilised just as effectively in each different mode of communication. The "digital communications" are analysed through the online data captured, which enables a nuanced understanding of how users interpret and use these spaces to understand and examine their health, their bodies, and their identities, through the methodologies of interviews and guided diarising. This enables a comprehensive understanding of the user's life worlds and does not see social media and other online environments as separate contexts, detached from other "offline" spheres of life. This research and the methodological approach is multi-sited, as social and cultural spheres always are. It provides an examination of broader online-offline dynamics, recognising digital platforms as being interwoven with the offline world. The following sections will provide a detailed insight of the authors research experiences from conducting a mixed-methods approach to analyse digital food cultures, strengths and weaknesses of these methodologies, lessons learnt, and guidance on how to conduct your own research using such an approach.

Semi-structured interviews

Given the personal and sensitive nature of the health-related content, personal reflections, practices, and representations of the body under analysis, establishing trust between researcher and participant is a key consideration for any research exploring digital food cultures. Any geographical limitations (in my study, the participants were located abroad and throughout the UK) can be overcome through interviewing over Skype; "a telecommunications application software product for internet video chat and voice calls" (Skype.com, 2018). Due to the temporal nature of conducting research into self-tracking and social media users over time, regular contact is key and can be kept via email or telephone and enables the building of and establishing trusting relationships between participant and researcher (Karl, 2009). It is important to maintain a professional distance, but not to lessen a rapport, which might make participants inhibited and unable to relax when engaging with the researcher. This can be a particular challenge when discussing personal health issues, diet, or disease. In turn, supportive strategies should be adopted by the researcher as a means of demonstrating empathy towards the participants, as opposed to providing support or advice. The strategies entailed offering sympathetic comments when participants discussed traumatic health-related events (such as diagnosis of disease or injury), without offering guidance or advice in suggestion of how these experiences could be handled or dealt with individually and personally. This should be an ongoing consideration, with the priority being to minimise any advice in regard to health management strategies, which may be perceived as advocacy from the researcher to the participant, while maintaining a warm and trusting environment.

Using semi-structured interviews provides a thematic guide to elicit responses to set questions, but also to explore individual understandings, interpretations, and engagement with dominant themes, which were informed and identified by the theoretical framework which informs the research questions. In my research, the key themes focused on identifying users' conceptualisations of (ill/physical/mental) health, sharing, social media, self-tracking, morality, community, feedback, surveillance, and sociality. The researcher encouraged the participants to expand upon autobiographical reflections raised outside of the set questions, tailoring questions "in the moment" in response to comments made, which were relevant to the research questions, while drawing the focus back to the key themes and set questions. Holding a relatively short time period for interviewing (30–45 minutes) can dictate the return to the set questions if discussions shift off topic. This can also provide a reasonable time frame from which to steer the discussion, without cutting the participant off mid-flow. Tactfully steering the interview is a careful consideration when discussing health self-tracking, for the interviewer it is important to avoid abruptly changing the topic, as this may cause participants to perceive their response as uninteresting (or oversharing), causing embarrassment or inhibition, and potentially limiting future responses. This is an important consideration and interviewing tool to ensure speaking about such personal feelings around health and the body is carefully and empathetically managed

Role of multiple interviews

Having multiple interviews with the users in conjunction with the diary is key for depth of analysis of digital food cultures. The first interview can provide the participants an overview of the research project and to tease out initial reflections around health behaviours before undertaking the guided diary. Its purpose is to aid participants' thought processes and to encourage them to engage with or challenge their own assumptions or interpretations of key themes and concepts with regards to health, food, and lifestyle management, which in turn aids the written dairies in terms of how perspectives may have changed prior to completing the diary. Introducing the diary, and briefly running through the guided questions also gives participants a space to clarify if any questions or prompts are unclear. The aim of the final interview is to contextualise diary entries (Meth, 2003) and draw together the narrative(s) that arose from completing the diary over time. This interview also enables both participant and researcher to explore topics outside of the guided diary questions and to discuss any written language which **may** have been misunderstood by the researcher.

An integral aspect of the final interview post completion of the diary was to explore topics, understandings, or interpretations, which arose from the participants' process of completing the diary over time. This enables an analysis of the participants' changing habits, practices, and broader shifting social media etiquettes. In turn, when working with multiple participants/users, once all have completed the diary, they can then be critically analysed together. For example, if thematic analysis is used to analyse the language use (from the written language in the diaries and the online data), this can be used to identify the key themes and questions to be addressed in the final interview. The final interview can then cover themes identified from analysing all the guided diaries and first interviews, as well as specific questions tailored to each participant's individual diary entries. It also provides a deeper understanding for the researcher of the significance of the events or observations made by the participants in their diaries and everyday lives. This aids the analytical process, allowing the contextualisation of the individual accounts within a broader socio-economic, political, and cultural context. In the context of exploring users of self-tracking technologies and social media, interviews will identify health choices while drawing attention to individual feelings, decision-making, and subsequent behaviours of the users, which enables the identification and in-depth analysis of individual health practices. Interviews are an important contextualising method due to the at times de-contextualising nature of the guided diaries (Meth, 2003).

Guided diaries

Solicited diaries have been used in health research since the early 1930s (Burman, 1995). In media and digital cultures research, however, they are an underused method. Traditionally, solicited diaries have not often included written questions and prompts – what I term guided diaries – but have been used as a space to regularly record experiences, personal motives, feelings, and beliefs in an unobtrusive way (Jacelon & Imperio, 2005). They enable "autobiographical reflections about the participants' life worlds" (Kenten, 2010: n.p.), over a period of time. The utility of including a guided diary is to enable participant engagement with the process of using social media and self-tracking technologies, while providing access to the everyday ways in which users become aware of their health and food-related decisions, and how this related to their engagement and self-representation on platforms online. As Plummer (2001) and Kenten (2010) argue, diaries enable a focus on the aspects that shape participants' lives, rather than on the more visible or easily identifiable aspects of health behaviours, by providing a crucial link between the private mind and publicly shared content. In particular, the guided diaries were used to tease out the participants' everyday reflections and interpretations of daily health management, food cultures, and lifestyle. A second insight gained is that the diary interview method can benefit both the participant and the researcher by facilitating a slower-paced, reflexive style of research.

Guided diaries can enable the participants to reflect upon their self-tracking use, social media engagement, and sharing, "in the moment" or soon after the "event" (in relation to their decision-making process or sharing online). Interactive writing in this way is like having fieldnotes from participants (Andrusyszyn & Davie, 1997). This can provide detailed insights into decision-making processes and reflections which may have been otherwise impossible to answer and identify through interviews or viewing online content alone. The guided diaries, therefore, were constructed to elicit responses to set questions, with the participants also invited to expand upon topics not outlined. Kenten (2010: n.p.) describes how a diary can be designed "but the ways participants respond to these styles will vary." This research approach draws upon another strength of guided diaries in empowering the participant (Meth, 2003), as they become both observer and informant (Zimmerman & Weider, 1977). The temporal nature of completing these diaries multiple times (in my research this was twice a month over three months, totalling six entries), and each time answering the same questions, encourages participants to engage with understandings of how their health-related processes and practices have changed over time (if at all). This also enabled them to reconsider the same concepts, such as "health," sharing, self-tracking, community, surveillance, and sociality, to uncover for themselves how initial assumptions might be challenged or overturned by revisiting these same questions over a period of time. As Kenten (2010: n.p.) identifies: "the narratives produced (...) create access points into their everyday lives." The questions can be tailored to provide insights into different health and food-related experiences by directing the participant to focus their attention upon health decisions and social media behaviours, to increase their visibility and the significance of their everyday activities, which might be otherwise considered mundane or irrelevant.

The guided diaries were provided via email on encrypted A4 word documents (see below excerpt: Guided Diary Template). A key ethical issue to note here is for the researcher to ensure for data protection purposes that documents are sent and stored securely between researcher and participant.

TABLE 11.1 Excerpt: Guided diary template

ENTRY 1
DATE:
USAGE

1. **Did you use any self-tracking, health or lifestyle apps or social media platforms today? If so, for what, and why? If not, why not?**
(Forgot/did not want to/for consumption monitoring and/or exercise monitoring, sharing content/data, sharing lifestyle)

2. **How did you feel after usage/or once you had decided not to use the application/platform?**
(Happy, satisfied, proud, frustrated, guilty, annoyed, relaxed, etc.)

SELF AND COMMUNITY

3. **Did you share any content of your health behaviours/lifestyle today on any form of social media, and if not, why not?**

   - What were your motivations behind sharing or privatising your content/data?
   - How did this make you feel?
   - Did you get any feedback/responses (or lack thereof) from other users and how did this make you feel?

4. **Did you feel your health decisions today contributed to your sense of self?**

   (As a "healthy" gym goer for example, or having a day off exercise, or a food "cheat" day for pleasure or relaxation)

PROCESS
Can you identify what (if anything) influenced and motivated your health choices and lifestyle today? (What you chose to eat/drink, whether you did any exercise, or abstained from any foods/drink/exercise?)

Screen Shot of Online Content: **Date:**

The diary template acted as a guide, including questions and suggestive answers to help participants who may have been feeling overwhelmed or intimidated by the blank page. In essence, they acted as a form of encouragement. Participants can find completing the diary to be an insightful experience which can "deepen the understanding of the relationship between a person's own unique socio-cultural situation and their online social interactions and experiences" (Kozinets, 2015: 61). They perceived the temporal nature of engaging with and reflecting back upon their health-related behaviours and sharing practices as a tool to "understand themselves better." This also provided them with a chronological and guided diary of a period of their lives in which many of them experienced personal life changes or reached personal goals: from recovering from life-threatening surgeries, to relocating, to refining a specific skill such as achieving certain yoga poses, to completing their first marathon. While trying to avoid framing the benefits of taking part in this research as celebratory personal revelations, many of the participants reflected in the final interview on how this process, over such an extended period of time, encouraged new personal insights (about themselves) or lifestyle changes, (sometimes "for the better"). For example, some participants who struggled with their over-regulatory self-policing of individual health (once this was personally recognised) at times managed to resist the dictatorial frames imposed by themselves and these technologies. For others, it simply highlighted their "unhealthy" relationships with food (eating disorders) or low self-esteem and body image issues. Some simply found it frustratingly time-consuming, but thankfully for the researcher, continued with its completion as they had "committed." A limitation of this method, then, is the labour-intensive aspect, which relies upon participants to be self-motivated and inspired to notice and record their thoughts about their sharing, health, and food-related practices.

Diary reflections unearthed individual observations concerning health behaviours and perspectives upon the process of how and why participants used these technologies, how this made them feel, and if this influenced subsequent behaviours. The repetitive task of completing the diary every month enabled continuity, and a reflexive journey for participants in identifying how, over a period of time, health behaviours, practices, and online community engagement changed (if at all). For example, analysing user data from these devices, even in combination with interviews, would not have enabled the fascinating and detailed everyday perspectives and initial reactions to the use of self-tracking devices, content shared, or social media community (dis-)engagement or feedback. Interviews would have provided a discussion and reflection of these processes and relational aspects, but likely "some-time after the event," whereas guided diaries were completed by the participants on the same day as the content shared (on different days of the week over the three-month period). This provided a close reading, reflection, and analysis of the day's health "events," both for the participant and the researcher. Another drawback to the diary interview method is that, despite providing detailed guidance, some participants were not always sure what to record in their diaries, this is when the importance of consistent communication between researcher and participant was key to ensure clarification of questions and responses. Another potential limitation of using diaries for research purposes is the recognition that writing diaries for consumption, unlike documentation for personal recollection, may encourage participant self-censorship. However, in relation to other ethnographic participatory methods, Kenten (2010) argues that self-censorship is as easily enacted within interviews or focus groups, as all methods of data collection can be to an extent considered partial. Kenten (2010) and Verbrugge (1980) argue that diaries offer the potential to provide a more accurate account of memories or events, due to the regularity and relative immediacy of the recollection in minimising memory errors. Self-censorship, however, can become transparent and easily identifiable through the continuity enabled through analysis of written content, especially when undertaken over an extended period of time.

Online data capture – Screenshots

Texts produce forms of knowledge, which help individuals and societal groups to make sense of and construct their environment (Willig, 2001). Critically analysing multiple curated screenshots from the participants allows for the historicising of social practices. Analysing these texts in relation to the timeline of engagement, enables a reflection on how responses, practices, and behaviours, online and offline, changed over time. This enabled a further exploration of shifting social practices and their relationship with disciplinary (neoliberal) discourses of power and control, arguably advocated and encouraged through evolving social media use and techno-commercial affordances. Context collapse, which refers to the merging social networks participants have on different platforms, is a key consideration and allows participants to engage with individuals from different aspects of their lives (Marwick & Boyd, 2010), for example, what they shared on Instagram in opposition

FIGURE 11.1 Screenshot of participant's shared food content.

to Facebook, and why? Deliberations between what content goes on which platform was a key ongoing consideration for participants, especially when it came to sharing visual content representing their food intake and bodies, as this made often for "complex and unpredictable uses, reuses, trajectories and uptake" (Varis, 2014: 4). Unknown trajectories of participants' content, particularly fear of unexpected surveillance (being viewed by those outside of the participant's network), was a key individual determinant and technological mediation as to what types of visual content were uploaded and where. As Mason (2002: 108) identifies:

In a more interpretive sense, (...) documents, visual images, objects, visualisation, and so on need to be "read" and interpreted in the context of, for example, how they are produced, used, what meanings they have, what they are seen to be or to represent culturally speaking.

What was useful about obtaining screenshots of online content is that the guided diaries at times afforded the researcher knowledge of the participants' "intended" message, with respect to how they wanted such online and offline representations to be interpreted and how then these representations would be perceived by their social media communities. Additionally, this analysis also enables an investigation of speculative interpretations related to how the image may actually be received by the community (or not), and the social connotations related to the way it was curated. As Mason (2002: 105) argues, "Memories, dreams, thoughts, plans, may thus be visual but not visible. This directs our attention to how the visual is embedded in the social, how it works, how we work with it" calling for methodological creativity in our research approach and in how this methodology grasps the social, political, and cultural context of the data in the analysis. Therefore, the final interviews contextualised participant reflections on image construction and representations of health, food, and identity. This enabled an examination of how it was received over time, and in turn how this affected their sense of self and moulded future considerations of self-representational food-related content. The triangulation enables different readings of the same image in different contexts and through different analytical lenses, enabling interpretative flexibility, while remembering that texts (verbal, written, and visual) are always constructed (Mason, 2002).

Conclusion

This chapter has explored an innovative methodological approach to investigating online and offline gaps and negotiations by social media and self-tracking users, offering knowledge of how users mediate their health practices, and food choices, providing empirical research that have not yet been done within the field of digital food cultures. Interpretation is the most significant challenge and strength of ethnographic research, as there is no transparent way to represent the social world (Atkinson & Hammersley, 1994: 254). Therefore, from the different approach's ethnography can harness, there is no "single philosophical or theoretical orientation that can lay unique claim to a rationale for ethnography" (Atkinson & Hammersley, 1994: 254). This means that flexibility is at the fingertips of the ethnographer.

Interpreting culture fairly is a true skill as "human phenomena do not arrange themselves obligingly in types but, rather, afford us the spectacle of endless overlapping" (Henry, 1971/2004: xv–xi). Therefore, this research approach does not consider the supposed "challenges" of digital ethnography as anything "new" in terms of analysing cultural and social phenomena. As Hine (2013 [2005]: 9) argues, in line with the ethnographic commitment to reflexivity: "the question is much more interesting, potentially, than whether old methods can be adapted to fit new technologies. New technologies might, rather, provide an opportunity for interrogating and understanding our methodological commitments." The real challenge and strength of ethnography is to confront the "(im) possibility of representing others" (Probyn, 1993: 61). This research approach negotiates these challenges by acknowledging the arbitrary nature of language and interpretations of behaviour. From the "back door" analysis achieved through the interviews and guided diaries, to the "front door" performativity of the curated online content (Goffman, 1959), these methodologies enabled an investigation of how participants' language use and health practices are shaped by political and socio-cultural conceptualisations of "health," as well as how they interpreted their own understandings of represented food consumption and related behaviours.

The participants all asserted that they were committed to tracking, monitoring, and sharing their health and food-related behaviours. Yet, life unsurprisingly at times prevented this cycle of engagement with these technologies. This resonates with Kristensen and Ruckenstein's (2018: 2) assertion that "self-trackers define and refine the limits and aims of self-tracking (...) everyday lives are characterised by volatile and less stable inter-relationality with self-tracking technologies which might be rejected, ignored, forgotten, tampered with, or used sporadically and irregularly." When life prevented these processes, the guided diaries enabled an examination of practices or inactivity related to health management and technological use, regardless of what the participants were able to "achieve" in their daily lives. The interviews allowed the participants to engage with broader and contextually informed ways of understanding what "health" and diet meant to them, why they made the decisions around food that they did, and how these practices and understandings shifted and evolved through changes in their personal lives. The temporal nature of this methodological approach then enables a broader examination of these practices, and in particular technologies' role in users lives over an extended period of time. These social practices constantly develop, overlap, and redefine themselves, and as such this innovative methodological approach enables insightful ongoing examination into the influences of digital food tracking through multiple digital and offline spheres in users' everyday lives.

Dos

- Historically contextualise your approach
- Think about what specific offline practices came before these online food cultures.

- Conduct pilot interviews, and if you have time a shortened guided diary period so you can prepare sufficiently for the main study.
- As with any research project, but especially when working with human participants, time management is key and prepare to be flexible.
- Be prepared for each stage of the ethnographic research process to take longer than you might plan, participants are not robots, and life disruptions happen.
- Check in with your participants to see how they are progressing with the diary; they will have ongoing questions.
- Be prepared to discuss your research question and design in lay person terms to your participants (and those outside of academia).
- Be prepared to adapt your guided diary questions after the first interview, the same goes for your final interview after participants have submitted their diary. This is a flexible, inductive, and adaptive research process.
- Build rapport with your participants and prepare for the rewarding benefits and research insights it brings!

Don'ts

- Leave recruitment to the last minute, it always takes longer than you think to get people on board.
- Feel disheartened when people drop out, life disruptions happen.
- Be available to your participants 24/7, set boundaries with your time and availability.
- Underestimate that working with human subjects means being flexible – delays will happen that are out of your control.
- Recruit too many participants. This methodological approach generates a wealth and depth of data. Think critically about what your research question needs in terms of recruitment numbers and demographic variety.
- Be disorganised with your data, ensure all interview, diary, and online data are securely stored on your device(s).
- Play down to your participants how much of their time the study will take. If you are not completely honest about how labour-intensive these approaches can be, they may drop out in the early stages of completing the diary. You will have wasted your time and theirs and will have to start recruiting again.

Now you try: Conduct a pilot digital food study

1. Identify three aspects of digital food cultures and practices that interest you, that would benefit from the long-term analysis and the in depth methodological approach of interviews, guided diaries, and online content. These could be practices you have witnessed within your social or professional network (both online and offline). For example, be related to personal health goals documented on digital platforms, or food related businesses marketed on social media.

2. Draw up your guided diary questions based on existing research and your interest in these digital food practices. (Use the template included in this chapter as a guide).
3. Recruit three participants – use your existing social and professional networks online and offline to promote your call. Ensure your call for participants is written in lay person terminology and language with limited academic jargon.
4. Based upon participant demographics and your research, draw up your semi-structured interview questions. Have set questions for all participants and some tailored on individual digital food and sharing practices.
5. Get participants to complete your pilot guided diary for three weeks, one entry per week on different days of the week, which include screenshots of their digital food-related content and practices.
6. Once completed, analyse all the diaries and first interviews together using your chosen method of analysis (e.g., discourse, thematic, textual analysis, etc.) based upon which is most appropriate for your research question.
7. Based on your analysis, draw up the final interview questions. These will be based on the overall findings across all diaries and interviews, and questions specifically tailored for each participants' individual insights.
8. Analyse final interviews, triangulating all the data you have collected and insights over the month-long interaction with your participants.
9. Share your research findings with your participants if they are interested to learn more – mine were!

Questions for reflection

What are the key insights you have gained from learning about these triangulated research methods?

Has this challenged existing notions you may have held about the role of and relationship between a researcher and participant in ethnographic studies? If so, how?

Reflect on your conceptualisation of digital food studies, is it the same as it was before reading this chapter? Think about the relationship between related online and offline practices in digital food cultures?

What most interests you in the field of digital food cultures?

Ethical considerations

Conducting (digital) ethnography presents many ethical considerations which must be adapted to the unique environment, interactive, and participatory spaces under examination.

Obtaining ethical approval is key. Consent forms must be obtained from participants, and the researcher must also provide information forms to make clear to participants their commitment to the study, and where the research will be published and for what purposes.

Ensure you clarify what you expect from the participants for such a long-term and in-depth study, they are kindly giving up their free time and sharing personal information which may make them feel vulnerable at points. Be grateful, transparent, supportive, and facilitate a conversation prior to consent to be clear about the goals of the study for researcher, and the role and potential impact of the participants self-analysis.

For data protection purposes, guided diary documents must be emailed in encrypted files/stored online securely between researcher and participant(s).

Given the sensitive nature of the subject matter (ill/health, disease, diet, lifestyle, etc.) and longevity of the study maintaining trust and regular contact is key between researcher and participant(s).

Balance maintaining supportive and empathetic strategies towards participants (e.g., in response to discussion of traumatic life events such as diagnosis of illness/disease), without offering guidance or advice. Ensure you maintain professional distance, but not to lessen a rapport, which might make participants inhibited when engaging with the researcher.

Consider the ethical issues at every step of the research process, from the planning, participant recruitment, to interviews, throughout the guided diary period, to analysis and dissemination.

Further readings

Cederström, C., & Spicer, A. (2015). *The Wellness Syndrome*. London: John Wiley & Sons.

Heyes, C. J. (2006). Foucault goes to weight watchers. *Hypatia*, 21, 126–149.

Wolf, G. (2010). The data-driven life. *New York Times*. April 28. Retrieved from www.nytimes.com: http://www.nytimes.com/2010/05/02/magazine/02self-measurement- t.html?_r=3&ref=magazine&pagewanted=al

References

Ajana, B. (2017). Digital health and the biopolitics of the Quantified Self. *Digital Health*, 3, 1–18

Ajana, B. (Ed.). (2018). *Metric Culture: Ontologies of Self-Tracking Practices*. London: Emerald Publishing.

Anderson, B. (1983). *Imagined Communities: Reflections on the Origin and Spread of Nationalism*. London: Verso.

Androutsopoulos, J. (2008). Potentials and limitations of discourse-centred online ethnography. *Language@Internet*, 5, 9–18.

Andrusyszyn, M., & Davie, L. (1997). Facilitating reflection through interactive journal writing in an online graduate course: A qualitative study. *Journal of Distance Education*. Retrieved from cade.athabascau.ca/vol12.1/andrusyszyndavie.html

Atkinson, P., & Hammersley, M. (1994). Ethnography and participant observation. In N. K. Denzin & Y. S. Lincoln (Eds.), *Handbook of Qualitative Research* (pp. 248–261). London: Sage.

Beato, G. (2012). The quantified self. *Reason*, 43 (8), 18–20.

Bottles, K. (2012). Will the quantified self movement take off in health care? *PEJ, the Physician Executive. Journal of Medical Management*, 38 (5), 74–75.

boyd, d. (2008). How can qualitative internet researchers define the boundaries of their projects: A response to Christine Hine. In A. N. Markham & K. N. Baum (Eds.), *Internet Inquiry: Conversations About Method* (pp. 26–32). Los Angeles: Sage.

Brewer, J. D. (2000). *Ethnography*. Buckingham: Open University Press.

Briggs, C. L. (1986). *Learning How to Ask. A Sociolinguistic Appraisal of the Role of the Interview in Social Science Research*. Cambridge: Cambridge University Press.

Brighenti, A. M. (2010). On territorology: Towards a general science of territory. *Theory, Culture & Society*, 27(1), 52–72.

Brin, D. (1998). *The Transparent Society: Will Technology Force Us to Choose between Privacy and Freedom?* Cambridge, MA: Perseus Book Group.

Bryman, A. (2016). *Social Research Methods*. Oxford: Oxford University Press.

Burman, M. (1995). Health diaries in nursing research and practice. *Journal of Nursing Scholarship*, 27(2), 147–152. doi: 10.1111/j.1547-5069.1995.tb00839

Burrel, J. (2009). The field site as a network: A strategy for locating ethnographic research. *Field Methods*, 21 (2), 181–199.

Butterfield, A. D. (2012). Ethnographic assessment of quantified self meetup groups. *QuantifiedSelf.com* Retrieved from http://quantifiedself.com/wpcontent/uploads/2012/05/Adam-Project-Report-5-17.pdf

Carmichael, A. (2010). Why I stopped tracking. *QuantifiedSelf.com*. Retrieved from http://quantifiedself.com/2010/04/why-i-stopped-tracking/11

Chen, J., Cade, J. E., & Allman-Farinelli, M. (2015). The most popular smartphone apps for weight loss: A quality assessment. *JMIR mHealth uHealth*, 3. Retrieved from http://mhealth.jmir.org/2015/4/e104/

Conor, L. (2004). *The Spectacular Modern Woman: Feminine Visibility in the 1920s*. Bloomington, Indiana: Indiana University Press.

Conquergood, D. (2009). Rethinking ethnography: Towards a critical cultural politics. *Communication Monographs*, 58 (2) 179–194.

Crawford, K., Lingel, J., & Karppi, T. (2015). Our metrics, ourselves: A hundred years of self-tracking from the weight scale to the wrist wearable device. *European journal of cultural studies*, 18 (4–5), 479–496.

Didžiokaitė, G., Saukko, P., & Greiffenhagen, C. (2018). Doing calories: The practices of dieting using calorie counting app MyFitnessPal. In B. Ajana (Ed.), *Metric Culture: Ontologies of Self-Tracking Practices* (pp. 137–155). London: Emerald Publishing Limited.

Flaherty, S.-J., McCarthy, M., Collins, A., & McAuliffe, F. (2018). Can existing mobile apps support healthier food purchasing behaviour? Content analysis of nutrition content, behaviour change theory and user quality integration. *Public Health Nutrition*, 21, 288–298.

Fotopoulou, A., & O'Riordan, K. (2016). Training to self-care: Fitness tracking, biopedagogy and the healthy consumer. *Health Sociology Review*, 26 (1), 54–68.

Fourcade, M., & Healy, K. (2017). Seeing like a market. *Socio-Economic Review*, 15(1), 9–29.

Garcia, A. C., Alecea, I., Standlee, J. B., & Yan, C. (2009). Ethnographic approaches to the internet and computer-mediated communication. *Journal of Contemporary Ethnography*, 38, 52–84.

Geertz, C. (1973). *The Interpretation of Cultures*. New York: Basic Books

Georgakopoulou, A. (2013). Small stories and identities analysis as a framework for the study of im/politeness-in-interaction. *Journal of Politeness Research*, 9(1), 55–74.

Gill, R. (2007). *Gender and the Media*. Cambridge: Polity Press.

Goffman, E. (1959). *The Presentation of Self in Everyday Life*. New York: Anchor.

Gray, A. (2003). *Research Practices for Cultural Studies*. London: Sage.

Gray, M. A. (1982). Inequalities in health. The black report: A summary and comment. *International Journal of Health Services*, 12 (3), 349–380.

Gregory, A. (2013) "Is our tech obsession making anorexia worse?" *The New Republic*. December 18. Retrieved from http://www.newrepublic.com/article/115969/smartphones-and-weight-loss-how-apps-can-make-eating-disorders-worse

Henry, R. (Ed.). (1971/2004). *Writing Systems: A Linguistic Approach*. Malden, MA: Blackwell.

Hine, C. (2000). *Virtual Ethnography*. London: Sage.

Hine, C. (2013). Virtual methods and the sociology of cyber-social-scientific knowledge. In C. Hine (Ed.), *Virtual Methods: Issues in Social Research on the Internet* (pp. 1-13). London: Bloomsbury.

Hine, C. (2015). *Ethnography for the Internet: Embedded, Embodied and Everyday*. London and New York: Bloomsbury.

Jacelon, C. S., & Imperio, K. (2005). Participant diaries as a source of data in research with older adults. *Qualitative Health Research*, 15(7): 991–997. doi: 10.1177/1049732305278603

Karl, I. (2009). Technology and women's lives: Queering media ethnography. *Reconstruction*, Special Issue on Fieldworld and Interdisciplinary Research, 9, n.p.

Kent, R. (2018). Social media and self-tracking: Representing the 'Health Self'. In B. Ajana (Ed.), *Self-Tracking*. London: Palgrave Macmillan. doi: 10.1007/978-3-319-65379-2_5

Kent, R. (2020a). Self-Tracking and digital food cultures: Surveillance and representation of the moral 'healthy' body. In D. Lupton & Z. Feldman (Eds.), *Digital Food Cultures* (pp. 19–34). London: Routledge.

Kent, R. (2020b). Self-tracking health over time: From the use of Instagram to perform optimal health to the protective shield of the digital detox. *Social Media + Society Special Issue: Studying Instagram Beyond Selfies*, 1–14. doi: 10.1177/2056305120940694

Kenten, C. (2010). Narrating oneself: Reflections on the use of solicited diaries with diary interviews. *Forum: Qualitative Social Research*, 11(2), n.p.

Kozinets, R. (2015). *Netnography: Redefined* (2nd ed. London and Thousand Oaks: Sage.

Kristensen, D. B., Lim, M., & Askegaard, S. (2016). Healthism in Denmark: State, market, and the search for a 'moral compass'. *Health*, 2.0 (5), 485–504.

Leppänen, S., Kytölä, S., Jousmäki, H., Peuronen, S., & Westinen, E. (2013). Entextualization and resemiotization as resources for (dis)identification in social media. *Tilburg Papers in Culture Studies, Paper* 57. Retrieved from https://www.tilburguniversity.edu/upload/a3d5524e-4413-4772- 9f96-9fe0ee714c6f_TPCS_57_Leppanen-etal.pdf. Last accessed: 01/05/18.

Levinson, C. A., Fewell, L., & Brosof, L. C. (2017). 'My Fitness Pal calorie tracker usage in the eating disorders'. *Eating Behaviours*, 27, 14–16.

Lewis, T. (2018). Digital food: From paddock to platform. *Communication Research and Practice*, 4(3), 212–228. doi: 10.1080/22041451.2018.1476795

Lupton, D. (2018). 'I just want it to be done, done, done!' Food tracking apps, affects, and agential capacities. *Multimodal Technologies and Interact*, 2, 29. doi: 10.3390/mti2020029

Lupton, D. (2019). Toward a more-than-human analysis of digital health: Inspirations from feminist new materialism. *Qualitative Health Research*, 29(14), 1998–2009. doi: 10.1177/1049732319833368

Lupton, D. (2020). Introduction: Understanding digital food cultures. In D. Lupton & Z. Feldman (Eds.), *Digital Food Cultures* (pp. 1–16). London: Routledge.

Lupton, D., & Feldman, Z. (Eds.). (2020). Introduction: Understanding digital food cultures. In *Digital Food Cultures*. London: Routledge.

Mackay, H. (2005). New connections, familiar settings: Issues in the ethnographic study of new media use at home. In C. Hine (Ed.), *Virtual Methods – Issues in Social Research on the Internet*. New York: Macmillan.

Magnet, S. (2011). *When Biometrics Fail: Gender, Race, and the Technology of Identity*. Durham, NC: Duke University Press.

Mann, S. (2005). Sousveillance and cyborglogs: A 30-year empirical voyage through ethical, legal, and policy issues. Presence: Teleoperators and virtual environments. *Health Sociology Review*, 14 (6), 625–646.

Marwick, A. (2015). Instafame: Luxury selfies in the attention economy. *Public Culture*, 27, 137–160. doi: 10.1215/08992363-2798379

Marwick, A. E., & Boyd, D. (2010). I Tweet honestly, I Tweet passionately: Twitter users, context collapse, and the imagined audience. *New Media & Society*, 13 (1), 114–133.

Mason, J. (2002). *Qualitative Researching*. London: Sage;

McRobbie, A. (2009). *The Aftermath of Feminism: Gender, Culture and Social Change*. London: Sage.

Mejova, Y., Abbar, S., & Haddadi, H. (2016). *Fetishizing food in digital age: #foodporn around the world*. In *Tenth International AAAI Conference on Web and Social Media* (ICWSM 2016) (pp. 250–258)Cologne: Association for the Advancement of Artificial Intelligence.

Meth, P. (2003), Entries and omissions: Using solicited diaries in geographical research. *Area*, 35, 195–205.

Middha, B. (2018). Everyday digital engagements: Using food selfies on Facebook to explore eating practices. *Communication Research and Practice*, 4(3), 291–306.

Moore, N. (1983). *How to Do Research*. London: Library Association.

Moore, P., & Robinson, A. (2016). The quantified self: What counts in the neoliberal workplace. *New Media & Society*, 18 (11), 2774–2792.

Moore, S. (1993). *Interpreting Audiences – The Ethnography of Media Consumption*. London: Sage.

Murphy, P. D., & Kraidy, M. M. (Eds.). (2003). *Global Media Studies – Ethnographic Perspectives*. London: Routledge.

Murthy, D. (2008). Digital ethnography: An examination of the use of new technologies for social research. *Sociology*, 42 (5), 837–855.

Neff, G., & Nafus, D. (2016). *Self-Tracking*. Cambridge, MA: MIT Press.

Parks, M. (2011). Social network sites as virtual communities. In Z. Papacharissi (Ed.), *The Networked Self: Identity, Community and Culture on Social Network Sites*. New York: Routledge.

Pink, S., & Fors, V. (2017). Being in a mediated world: Self-tracking and the mind-body-environment. *Cultural Geography*, 24, 375–388.

Plummer, K. (2001). *Documents of Life 2: An Invitation to a Critical Humanism*. London: Sage.

Postill, J., & Pink, S. (2012). Social media ethnography: The digital researcher in a messy web. *Media International Australia*, 145(1), 123–134.

Probyn, E. (1993). *Sexing the Self: Gendered Positions in Cultural Studies*. Oxford: Routledge.

Punch, K. (2014). *Introduction to Social Research: Quantitative & Qualitative Approaches*. London: Sage.

Rettberg, J.W. (2014). *Seeing Ourselves through Technology: How We Use Selfies, Blogs and Wearable Devices to See and Shape Ourselves*. Basingstoke: Palgrave Macmillan.

Robinson, L., & Schulz, J. (2009). New avenues for sociological inquiry: Evolving forms of ethnographic practice. *Sociology*, 43, 685–698.

Rousseau, S. (2012). *Food and Social Media: You Are What You Tweet*. Lanham, Maryland: Rowman Altamira.

Ruckenstein, M., & Kristensen, D. B. (2018). Co-evolving with self-tracking technologies. *New Media & Society*. doi: 10.1177/1461444818755650

Ruckenstein, M., & Schüll, N. D. (2017). The datafication of health. *Annual Review of Anthropology*, 46, 261–278.

Rymes, B. (2012). Recontextualising YouTube: From macro-micro to mass-mediated communicative repertoires. *Anthropology & Education Quarterly*, 43(2), 214–227.

Sade-Beck, L. (2004). Internet ethnography: Online and offline. *International Journal of Qualitative Methods*, 3(2). Retrieved from http://www.ualberta.ca/~iiqm/backissues/3_2/html/sadebeck.html. Last accessed: 12/07/19).

Simpson, C. C., & Mazzeo, S. E. (2017). Calorie counting and fitness tracking technology: Associations with eating disorder symptomatology. *Eating Behaviours*, 26, 89–92.

Varis, P. (2014) *'Digital Ethnography' Tilburg Papers in Cultural Studies 104*. Retrieved from https://www.tilburguniversity.edu/upload/c428e18c-935f-4d12-8afb- 652e19899a30_TPCS_104_Varis.pdf

Verbrugge, L. (1980). Sex differences in complaints and diagnoses. *Journal of Behavioural Medicine*, 3(4), 327–355.

Willig, C. (2001). *Introducing Qualitative Research in Psychology: Adventures on Theory and Method*. Buckingham and Philadelphia: Open University Press.

Zimmerman, D., & Weider, D. L. (1977). Diary-interview method. *Urban Life*, 5(4), 479–498.

Zion Market Research. (2018). Global MHealth Apps. *Globalnewswire.com*. Retrieved from https://www.globenewswire.com/news-release/2019/01/24/1704860/0/en/Global-mHealth-Apps-Market-Will-Reach-USD-111-1-Billion-By-2025-Zion-Market-Research.html. Last accessed: 08/12/20.

12

HOW TO USE DIGITAL MEDIA IN FOOD-RELATED ACTION RESEARCH

Jonatan Leer

The ambition of quite a lot of research within food studies is – more or less explicitly – to stimulate change. However, this goal takes many forms, for instance, critical discourse analysis of food politics seeking to question the current state of affairs (Warrer & Leer, 2018) or in follow-up studies of interventions which evaluate new initiatives (Andersen et al., 2017). The most overt form of research with an ambition to change things is action research (AR) (Reason & Bradbury, 2008). In this type of research, the researcher stimulates change in a collaborative process with practitioners. This approach to research has, nonetheless, a series of inherent challenges. Notably, it reframes a) the traditional role of the researcher; b) the relation between practitioners and researchers; c) the scientific and practical objectives of doing research.

In this chapter, I will discuss AR as a method within digital food studies and make a case for the usefulness of digital data generation in food-related AR projects. I will argue that the digital filming of practices via tablets and smartphones offers a way of securing the core values of this methodological approach, notably the inclusion of practitioners in both the collection and analyses of data. Also, I consider it important to highlight how different kinds of new knowledge can emerge from this empirical co-creation and, notably, how digital media formats like social media clips and podcasts can help disseminate the results beyond an academic readership such as politicians and practitioners.

I begin by briefly outlining the core ideas and struggles in the distinct interpretations of AR and I opt for the approach called critical utopian action research (CUAR) (Nielsen & Nielsen, 2005). In the next section, I will provide some examples of how AR has been applied within food studies and highlight the fact that the inclusion of digital media is remarkably rare in these projects. On this basis, I will present a case from my own research. The case concerns meal practices in day-care institutions in Denmark and how to make them more participatory. I will discuss how we framed and used AR in this project and particularly how digital data generation proved a central means to give practitioners a voice. Finally, I will give some thought to what I learned in the process and what I might alter in future explorations. I will also describe how the results of the project were disseminated via digital media and more traditional research channels.

Action research and food studies

Action research: Ambition and criticism

> We have described action research as "a family of approaches," a family which sometimes argues and falls out, whose members may at times ignore or wish to dominate others, yet a family which sees itself as different from other researchers, and is certainly willing to pull together in the face of criticism or hostility from supposedly "objective" ways of doing research.
>
> (Reason & Bradbury, 2008: 7)

AR was introduced in the mid-twentieth century as a critique of the dominant ideals of social research. People like Kurt Lewin, one of the initiators of the movement, felt that traditional research was too distant from the real world and its problems (Lewin, 1946). Rather than contributing to seek solutions to real and present social challenges, traditional researchers – according to Lewin and likeminded people – approached the world theoretically and often with a clear sense of authority and superiority to the subjects and the objects they studied. The ambition of AR was, through context-specific interventions, to answer concrete questions relevant to the citizens in question. Also, the researcher should not just observe and evaluate projects and write a report on how that might be done differently. The researcher should work with the people who are part of the context being studied. The ambition is to collectively find durable solutions. Thus, the core values in the knowledge generation process in AR are collaboration, democracy, sharing, experimentation, relevance, and applicability.

While all these values might have positive connotations to most democratically engaged researchers, they also challenge some fundamental assumptions in scientific practice. In particular, natural science values including objectivity, reproducibility, distance, and universalism and the more quantitative research are lost in AR. This has led to a severe criticism of AR from researchers who insist on these values. As we can see in the above quote, the AR community almost takes pride in this dispute and unites around this positivist anti-model. Also, other researchers who refute the positivist/objectivist approach to research are critical of AR. A central disapproval has been to problematise the role of the researcher as simultaneously activist, generator of new knowledge, and critical thinker (Nørholm & Petersen, 2002).

The role of the researcher is also heavily debated within the AR community. Some believe that anyone can do AR and refuse to give the academic researcher any privileged position in the process of generating and analysing data (McNiff, 2017). Others believe that distinct roles should be maintained in the process (Nielsen & Nielsen, 2005). Another central discussion concerns outcomes and the goals of AR. To McNiff, the actual change in local contexts is a primary concern. The publication of academic papers is less important. Others argue that the publication of academic works and works of a more popular and accessible nature are of great importance to sustain the change ambitioned by the project and to make it relevant to others (Nielsen & Nielsen, 2005).

In the same vein, the scope of knowledge generated by an AR project is a core point of debate. Some believe that AR project results are primarily of relevance in local contexts. Others believe that it is vital to connect each AR project to major societal and scientific debates. A generally applied distinction is the one between first person (individual level), second person (the "we" of a specific case), and third person (the ability to create a wider, societal impact and theory) (Torbert & Taylor, 2008).

A more recent criticism of AR is that it no longer fulfils the democratic and subversive goal of social change originally envisioned. Contemporary research policies increasingly embrace a neoliberal logic where the value of research is manifest in its applicability and orientation towards rapid solutions rather than in its scientific depth and originality. In this context, some see AR as an instrument to implement this particular political agenda and, as such, it distances researchers from critical thinking and more fundamental questions about knowledge production (Johansen et al., 2018).

My approach to action research: A "polygamist" researcher's view

Academically, I was brought up as far removed from AR as one can be in academia, namely literary studies. In this tradition, we study texts. Mostly, we used an advanced French theorist to apply a new angle to a canonical book and provide an even more abstract interpretation than previous readings. These essays had to be written in a very sophisticated manner, so that only our peers schooled in the same manner could understand them. When I started to work with food, I was slowly dragged away from my desk and into the "actual" world. I still used my analytical skills and my French theorists, but the objective of my work came closer to everyday life and real people. Also, I began to understand that people other than my peers were interested in my research and I was invited to talk and write about my research in various popular and academic formats.

This journey – probably culminating with the AR project discussed in the next section – has made me think about different modes of research and the distinct potentials of these forms of knowledge generation as well as the forms of communicating this new knowledge. I have a lot of sympathy for the ambitions of AR, but I foresee much legitimate criticism of this approach. So, I do not cultivate a "monogamous" relationship with AR. I do believe that other forms of research are necessary – for me, too, and my identity as a researcher. I feel, however, that the AR projects I have been involved in are maybe some of the most meaningful parts of my research career.

This means that I am not dogmatic in respect of AR. I use a range of methods and forms of research. When I engage in AR, I consider it important not just to assume the role of consultant. It makes sense to aspire to use research as a means to better concrete contextual challenges. Nevertheless, we should think hard and critically in a collective manner about how the case(s) can contribute to wider societal and scientific debates. Thus, my position is close to that of CUAR. This is a particular Nordic tradition of AR, emphasising a reflective, dialectic process in which perspectives based on everyday practice and theory meet and continuously challenge each other (Nielsen & Nielsen, 2005, 2010, 2016).

This tradition is rooted in critical theory. As a school of thought, critical theory is often associated with thinkers like Adorno. It was committed to criticising cultural hegemony and supporting suppressed individuals. Also, the ambition was to formulate new, alternative models of social interaction and society. This work involved mostly textual analysis of popular culture and philosophical arguments. As such, the earliest works within critical theory had little connection to the everyday life of suppressed people. In CUAR, Nielsen and Nielsen seek to use the analytical perspective from critical theory and combine it with AR's ambition to develop actual change locally in collaboration with the people concerned. This change is often generated by challenging the status quo via collective experiments of alternative futures and forms of working together. So, contrary to certain forms of AR, CUAR is balancing between practice and theory. On the one hand, the researcher is committed to changing things in a local context, but, on the other hand, also to research and to generating critical thinking by connecting the local perspective to a more theoretical or general level of thought and critique. This means that CUAR has a strong focus on the responsibility of the researcher. Hence, although democracy and collective thinking are central to CUAR, the researcher retains a specific position. The end goal of CUAR is thus not merely to provide change locally as in certain forms of AR, but also essentially to formulate a critical reflection on the basis of this change-giving process.

Examples of action research in food studies

AR is highly applicable when rethinking social organisation and/or stimulating change in underprivileged contexts and, notably, in relation to currently predominant issues in food studies: sustainability, food justice, health inequality, food education, etc. Nonetheless, it is actually rather difficult to find literature in food studies using an AR approach. A search in major journals in the field, such as *Food culture and Society*, *Gastronomica*, and *Food and Foodways*, does not provide a single title.

Yet, I managed to find a few studies involving projects on improving food practices in a diverse range of contexts: sustainable food systems in Vancouver (Rojas et al., 2017) and Cardiff (Moragues-Faus et al., 2016), food distribution to the elderly in Sweden (Pajalic et al., 2012), local organic farming (Helmfrid et al., 2008), smart rural development (Soulard & Lardon, 2019), women's agency for food security (Williams, 2014), and digital organisation as a tool for food revolution for precarious youth (Akom et al., 2016).

As this overview demonstrates, these studies cover a range of topics, but there seems to be a general concern with either institutional change or with improving food for marginalised groups (precarious youth, women, farmers, the elderly). Also, the approaches differ and the research designs are quite distinct as are their understandings of AR. Most include a clear participatory focus. Also, the project is designed in various steps, often with a circular structure. This circular structure secures a dialectic relationship between defining problems, hearing parties, taking new action steps, reflecting on/evaluating the experiments, and redefining the case on the basis of new knowledge. One of the most elaborate designs is found in Helmfrid et al. (2008) with a multitude of interdependent steps, experimental reflection sessions, and new action initiatives, evaluations, etc. I could only find one study that involved the use of digital media, namely Akom et al. (2016). In this study, young people used digital media to document and map food inequality in their neighbourhoods, thus beginning to change the food system.

I will now turn to my case where digital media was used to develop the meals in day-care institutions in Denmark.

The case – Day-care professionals

From 2015 to 2018, I was involved in an AR project called Children, Food and Meals[1]. The project was commissioned by The Danish Veterinary and Food Administration (DVFA). The goal was to improve children's eating experience in Danish day-care institutions. The project marked a change in the organisation's perspective on children and food. Previously, DVFA had been dominated by an exclusive focus on health and nutrition. What critics might call a nutritionist gaze (Scrinis, 2013). This present project had been formulated after a dialogue with Danish food researchers – including myself – from the Danish School of Education (DPU). At this institution, we were very concerned with formulating a new understanding of food, education, and children based on democratic, interactive, and contextualised understandings of food and children (Leer & Wistoft, 2018). This approach was rooted in the Nordic school of health pedagogy (Jensen, 2009) and was part of a more general global turn towards critical food pedagogies (Hayes-Conroy & Hayes-Conroy, 2013; Flowers & Swan, 2016). Although this turn takes many forms, it is fair to say that it is a criticism of "hegemonic nutritionism," an ideology that understands food in standardised, reductionist, decontextualised, and hierarchical ways (Hayes-Conroy, 2013: 1–4). Also, in the critical food pedagogies, food is understood locally as a part of political (often institutionalised) power structures. Food education should encourage agency and participation from the students and enable students to see and criticise the power structures of food systems as well as food inequalities related to race, class, gender, etc. (Flowers & Swann, 2016).

Traditionally, DVFA has insisted on the "hegemonic nutrition" focus on the standardised understanding of meals in day-care institutions, but with Food, Meals and Children, they integrated a more "critical-food-pedagogical" approach into their demand for research. They wanted to focus on how one could better children's experience of meals, not merely how you could get them to eat in accordance with the hegemonic nutritionist guidelines. Also, another ambition was to stimulate the collaboration between pedagogical personnel, management, and kitchen staff in order to develop pedagogical professionalism in respect of food to encourage children's agency. So, paradoxically, the project is a top-down project, however, with a clear ambition of stimulating bottom-up agency. The participatory AR approach was also chosen as a way of ensuring the bottom-up dimension.

Project design

The ambition of the project was to bring practitioners from a variety of day-care institutions together in an experimental setting to discuss their practices concerning meals. The goal was to facilitate new, more participatory approaches. Ninety-eight practitioners from 23 institutions across Denmark signed up for the project. Typically, most institutions would send several staff members from different professional categories: pedagogical staff, management, and kitchen staff. The focus on critical food pedagogies was based around three labs: one on experimental learning through food, one on interdisciplinary collaboration around meals, and one on food and taste pedagogies. I was responsible for the latter and I will focus on that setting for the rest of this chapter.

In the lab for food and taste pedagogies, the goal was to develop new practices which engaged the children in meals and stimulated their agency. Also, there was a strong ambition to stop seeing meals as a break from pedagogical activities, but rather as space and time for learning and competency development. The learning could entail social and intercultural competencies as well as learning about the biological and sensory qualities of the food or the cultural and historical aspects of the food. We were a total of 28 participants from all the professional categories mentioned and they came from five very different institutions. The fact that our starting point was not a local context, but five institutions in different contexts meant that this AR project stood apart from many others. Some day-care institutions were situated in urban areas while others came from rural areas. The use of digital media was a means to unite these different contexts. We met monthly for 11 sessions at DPU over the course of a year. Initially, the different institutions were asked to clarify a focus within the lab theme. For instance, how can we stimulate children's agency during meals? How can we include the children in the cooking activities? Or how can we incorporate our institutions' science and technology profile into the meals? They were then asked to find a way to document the current meal practices by filming it using a smart phone or tablet for the next session. This was the baseline documentation. In the subsequent session, they presented their digital documentation for all participants in the lab. Based on the collective discussion, which often sparked many suggestions, the group formulated new initiatives to stimulate the change they wanted to see. The results of these initiative were documented via digital filming. This digital material was then discussed in the next session, etc.

This circular form is very common in AR. Rather than providing rapid change, much AR values a long process with room for experiment and the assurance that all parties are heard. My role was first and foremost to facilitate this circulation of knowledge and ideas in the change-giving process. In the research team, we also talked about our role as "disrupting" the practitioners by asking questions and introducing the critical food pedagogy perspectives. The aim was not to tell them what to do or what to think, but to disrupt their doings and sayings about food. Nevertheless, we insisted that they were the experts concerning their practice and only they could see if the perspectives presented by us made sense to their specific context, and how the different theories could be adapted to this. Also, I saw it as my job to ensure that the knowledge generated in the labs was documented.

Usefulness of digital media

In the lab, the digital filming of meal practices was an eye-opener for most participants. Despite their differences, all participants had an ambition to develop more inclusive meal practices and had strong ideals concerning democratic and interactive meals. Andersen et al. (2017) have created a model which is very useful when describing different approaches to food pedagogies.

The model can be utilised to describe any kind of pedagogical food practice. In the project, the mealtime situation should be understood as a pedagogical activity and not just as a refuelling break. The vertical axe describes how the food pedagogues approach the children (either in an authoritative or more democratic manner or, most commonly, somewhere in between) and the horizontal axe describes the role of the food (as a means to an end or as having an intrinsic value, or, most commonly, somewhere in between). The top right corner, the authoritative and instrumental approach, leaves little space for children's agency while the bottom left corner leaves ample space for children's agency. Conversely, the role of the pedagogue is to dominate in the top right corner and be much less important in the bottom left corner. It is important that the model does not merely allow four distinct positions, the scheme should be seen as a continuum with a multitude of positions. Generally, the dominant pedagogical ideal in Danish schools and day-care institutions tends towards the bottom left. This was also the ambition of the practitioners in my lab. They wanted increasingly to remain in the background and let the children be more active in determining how much, what, and in which order they wanted to eat. This is a relatively new approach. For many years, children's eating has been viewed as a high-risk area in need of intense adult supervision and control (Leer & Wistoft, 2018). The practitioners in the lab felt that they were genuinely making space for children's agency.

We discussed the matter in the first session and decided that all the groups should film a meal session using their institutional iPads. In the next session, we watched and discussed the digital film clips. During this session, it became apparent that, in many cases, the adults did not really provide space for the children's agency to the extent perceived. I will focus on one particular team from one of the institutions. In this institution, an adult would sit down during mealtimes with a group of six children aged between three and five. In the film clips, the adults often performed a very active role in the meal situation. Distributing food, pouring water, policing what the children would put on their plate, and making sure that they would eat it in the right order. The children were accustomed to this and the adult policing and dominance left them passive. The children's agency was mostly expressed through rebellious behaviour, particularly from the older children.

When we watched these film clips collectively in the lab, it became apparent that although the adults in the situation had strong ambitions about making space for children's agency, the method used to organise the meals was not facilitating these ideals. So, we decided to work on the role of the adults and tried gradually to push the adults into the background. In the next session, they documented how they had worked on letting the adults stand behind the children. This was still not working very well. The film clip showed the hands of adults in the background constantly intervening in the children's actions, resulting in the children being very confused. The pedagogues were unaware of this at the time, because so much was going on. It was only during the collective reflection on the digital documentation in the lab that this became obvious to them. After several additional experiments, the children were finally divided into smaller groups. Three of the oldest sat alone at one table and the younger ones sat with an adult, who now acted more as someone who ate with the children and not as a host or supervisor. In the lab, we watched a clip of how this worked for the three older children on their own. In the film clip, they ate the classic Danish dish boller i karry (curried meatballs) with toasted coconut flakes and raisins as a topping. Not quite in keeping with tradition, the three children started by passing the topping around, as this was new to them. They then turned to the rice, the meatballs, and the sauce. To the surprise of most, the children ate everything and more than they normally did; also, they attentively passed the different foods around and made sure everyone got their fair share. The absence of adults made them responsible for both the eating and for each other. Luckily, the adults stayed away completely, also when they started eating the food in the wrong order by commencing with the toppings. Had the adults been at the table, "the wrong order" would most likely have been sanctioned.

To me this was a very enlightening example of how much space and agency children are able to handle if they are organised in small groups and given genuine responsibility. This was a surprise to most of the practitioners and inspired many to experiment in similar ways. I could use this example to challenge the research that is very sceptical about children's agency in relation to food (as described in Leer & Wistoft, 2018).

This example demonstrates how using documentation of practices via iPads worked to facilitate change through experimentation. If the adults had merely described what they did, I do not think they would have realised where things went wrong. They felt that they had the right vision and that their actions helped to realise their ideals. It was the digital documentation of their practice and the subsequent analysis and discussion of the digital data in the lab over a long period of time that inspired them to initiate change and to document it. In doing so, they also inspired others in the lab to experiment in similar ways. An important element here is that I did not tell them what to do. This was a collective reflection in the lab, based on circular dialogue – and several such dialogues – between practice and theory.

Also, it is important to emphasise that this process sparked change at various levels. It generated change locally in the institutions. It also inspired the other institutions participating in the project to experiment with similar approaches. Based on these experiments and the extended documentation, I was able to use the material to advance the scientific debate. In our final project report (Hansen et al., 2018), I used this documentation to challenge the more traditional approaches/hegemonic nutritionist approaches to meals in day-care institutions. This convinced the DVFA to change the official guidelines in this area, which might help to extend it on a much wider scale.

During the project, digital media played a central role in the dissemination of results. We would like to go beyond the report format which rarely attracts great audiences. We prepared a scientific report as mentioned, but we condensed the key points into a series of films and podcasts shared on the digital media – notably in social media for day-care professionals where they were the subject of extensive debate. In this digital dissemination of the project, we made sure not only to include the voices of the researchers, but also those of the practitioners.[2]

Lessons learned

I believe this example makes a strong case for using digital media as a tool in AR food projects. It can be particularly useful to show what is actually done. If you restrict your documentation to interviews, you might risk a response limited to what the practitioners believe they do. Also, the new electronic devices are part of everyday life in many social contexts. In Danish day-care institutions, for instance, tablets and smartphones are used to register the children and their locations, communication with parents, etc. They do not, therefore, attract the children's attention in the sense that something extraordinary is going on.

Furthermore, I felt that this way of documenting practices offered a voice to some of the people, who might otherwise have had some difficulty in formulating such observations in writing. One of the most important advantages of this kind of data is that it allows for collective reflection in the lab. Everyone got quite a clear sense of the spaces, contexts, and difficulties at the other institutions with many more nuances than I believe written statements would have provided. In the participants' evaluations, these collective sessions were highlighted as particularly meaningful, as they provided new perspectives and insights and opened for reflection on the particular practice of each participant. The digital films were central to this in-depth treatment and to really appreciating differences and similarities. They also helped to document innovative experiments for later discussion.

Nevertheless, I think we should always keep in mind that a digital film, like individual field notes, reflects a certain point of view. It is thus a subjective description and not the whole truth. For instance, I regret that the project focused almost solely on the adults' point of view. This was, of course, due to the way the project was framed, but I believe it could have been very interesting to include the children as generators of data. It would have been interesting to let the children handle the digital devices during a meal to see what they would focus on. We could, therefore, have used the digital media even more diversely to get more perspectives. I thought about this too late in the process to change the scientific design of the project.

This also points to certain ambivalences in this project and most types of AR. We like to make AR as inclusive and democratic as possible, although some hierarchal differences persist. In this case, it was the DVFA and, to some extent, the researchers who framed the project and had the authority to define what the project should focus on and how. Nevertheless, the researchers were the lab leaders and responsible for the summing up. So, although I tried to step down as much as possible in the lab sessions, I was the leader and the one who would sum up our work in the sessions and in the final report. Although I tried my best to include as many voices as possible and show my report to the practitioners for comments before submitting it, my conclusions certainly did favour certain voices rather than others and the report would most likely have been different, had it been written by another researcher. Also, in the data generation, there remained, as mentioned, a hierarchy between the children and the pedagogues. I think we should work hard to be as democratic as possible and open to as many voices as possible when undertaking AR, but I do not think that we can completely avoid these hierarchies. It is necessary to have a certain distribution of roles (and thus also a hierarchy) in order to complete such a project. However, I think it is important to be clear about this and to be mutually very respectful.

On the technical side, we experienced some technical issues with the digital devices during some sessions. Also, it could be rather complicated to share data and transferring them from a device to a safe platform where the data could be kept. We had permission to use the videos in the project, but we had to keep them on a secured platform (Dropbox, for instance, was not sufficiently secure) on the university website, which did not work and was difficult to access for the practitioners. This was a rather time-consuming and sometimes frustrating task. It would be nice, in the future, to have a more accessible, safe, and functional solution which everyone could access. Paradoxically, this can actually be rather complicated in the digital age.

Dos and Don'ts

Do have a clear plan for what your ideal outcomes should be in terms of balancing concrete change in a local context and research output with more generalisable perspectives.

Do reflect thoroughly about the kind of digital method that might be used productively in the project (tablet and smartphone filming is just one of many options), (see Akom et al. 2016)? and who can/should generate the data? How should the data set be managed and who should analyse it?

Do a list of all the different stakeholders/participants and try to understand their agendas and how these might conflict.

Don't act the brilliant researcher who has all the answers. It is a collective process.

Don't think it will be easy. A democratic, inclusive project is always very complicated and many unforeseen things will come into play.

Now you try

Pick a context (a local community, an institution, etc.) with a food-related problem (food inequality, food access, unhealthy eating patterns, poor infrastructures, etc.). Try to think about how you might approach this problem from a digital AR angle. Draw up a plan including a) a list of participants and different interests; b) a description of the digital media methodology and an argument for its usefulness; c) a timeline including the circular approach of meetings, field work, evaluations, changes, field work, and evaluation, etc.; d) a list of the outcomes and an argument for choosing these; e) a reflection on ethical issues and advantages and drawbacks of this method compared to a more traditional research and non-digital AR design. Make sure the plan is realistic and that there is a logical alignment between the different elements: the research question, the method, the participants, and the outcomes.

Questions for reflection

When could AR in digital food studies be very relevant and less relevant as a methodological approach? Give two examples.

What do you consider important as a researcher involved in an AR project in digital food studies? And what might be difficult or problematic?

Which other digital media, methods, or resources might be used in a food-related AR project?

Ethical considerations

Remember to obtain consent from all involved and to obtain parental consent for minors.

Save the data in a safe space and only keep the data for as long you need it.

If preparing collective analyses (as I would recommend), make sure to create a positive and including space where everybody feels accepted and any critical comments or suggestions are formulated in a constructive manner. Make a set of rules, for instance, which you present at the beginning.

Further readings

Leer, J. & Wistoft, K. (2018). Taste in food education: A critical review essay. *Food and Foodways*, 26(4): 329–349.

Nielsen, B. S. & Nielsen, K. A. (2016). Critical Utopian action research: The potentials of action research in the democratization of society. In H. P. Hansen et al. (eds.), *Commons, Sustainability, Democratization* (pp. 90–120). London: Routledge.

Reason, P. & Bradbury, H. (eds.) (2008). *Action Research: Participative Inquiry and Practice*. London: Sage.

Notes

1 https://projekter.au.dk/boern-mad-og-maaltider/).

2 https://www.facebook.com/watch/?v=482764852262099, https://www.youtube.com/watch?v=LD7oAH2T1Qo, and https://bupl.dk/artikel/ny-podcast-glaede-og-frihed-paa-menuen/)

References

Akom, A., Shah, A., Nakai, A., & Cruz, T. (2016). Youth Participatory Action Research (YPAR) 2.0: How technological innovation and digital organizing sparked a food revolution in East Oakland. *International Journal of Qualitative Studies in Education*, 29(10): 1287–1307.

Andersen, S. S., Baarts, C., & Holm, L. (2017). Contrasting approaches to food education and school meals. *Food, Culture & Society*, 20(4): 609–629. doi: 10.1080/15528014.2017.1357948

Flowers, R. & Swan, E. (eds.) (2016). *Food Pedagogies*. London: Routledge. doi: 10.4324/9781315582689

Hansen, O. H., Leer, J., Broström, S., Warrer, S. D., & Jensen, T. M. (2018). *Professionalisering og øget tværfaglighed i samarbejdet omkring mad og måltider i dagtilbud*. Tjele: DCA-Nationalt Center for Fødevarer og Jordbrug.

Hayes-Conroy, A. (ed.) (2013). *Doing Nutrition Differently: Critical Approaches to Diet and Dietary Intervention*. London: Routledge. doi: 10.4324/9781315577913

Helmfrid, H., Haden, A., & Ljung, M. (2008). The role of action research (AR) in environmental research: Learning from a local organic food and farming research project. *Systemic Practice and Action Research*, 21(2): 105–131.

Jensen, J. M. (2009). Hverdagsliv og sundhedspædagogik: konturen af en pædagogik der er sensitiv over for det levede liv. In V. Simovska et al. (eds.), *Sundhedspædagogik og sundhedsfremme: teori, forskning og praksis* (pp. 63–81). Århus: Aarhus Universitetsforlag.

Johansen, M. B., Schmidt, L. S. K., & Andersen, P. Ø. (2018). Aktionsforskning og kritikkens betingelser: når forskningsanvendelse bliver en selvfølge. *Dansk Pædagogisk Tidsskrift*, (1): 6–15.

Leer, J. & Wistoft, K. (2018). Taste in food education: A critical review essay. *Food and Foodways*, 26(4): 329–349.

Lewin, K. (1946). Action research and minority problems. *Journal of Social Issues*, 2(4): 34–46. doi: 10.1111/j.1540-4560.1946.tb02295.x

McNiff, J. (2017). *Action Research: Principles and Practice*. London: Routledge.

Moragues-Faus, A., Omar, A., & Wang, J. (2016). *Participatory action research with local communities: Transforming our food system*. Food Research Collaboration Review Paper, London. http://foodresearch.org.uk/wp-content/uploads/2015/11/FINAL-Participatory-Action-Research-with-Local-Communitiesreport-23-11-15-1.pdf (accessed 4.11.19).

Nielsen, B. S. & Nielsen, K. A. (2016). Critical Utopian action research: The potentials of action research in the democratization of society. In H. P. Hansen et al. (eds.), *Commons, Sustainability, Democratization* (pp. 74–106). New York and London: Routledge.

Nielsen, K. A. & Nielsen, B. S. (2005). Kritisk Utopisk Aktionsforskning: Demokratisk naturforvaltning som kollektiv dannelsesproces. In T. B. Jensen & G. Christensen (eds.), *Psykologiske og pædagogiske metoder: Kvalitative og kvantitative forskningsmetoder i praksis* (pp. 155–181). Roskilde Universitetsforlag: Roskilde.

Nielsen, K. A. & Nielsen, B. S. (2010). Aktionsforskning. In S. Brinkmann & L. Tanggaard, (eds.), *Kvalitative Metoder: En grundbog* (pp. 97–120). Copenhagen: Hans Reitzels Forlag.

Nørholm, M. & Petersen, K. A. (2002). *Videnskab og engagement: Staf Callewaert 70 år, den 16. juni 2002. [København]: Hexis – Forum for Samfundsvidenskabelig Forskning.* Viborg: PUC.

Pajalic, Z., Westergren, A., Persson, L. & Skovdahl, K. (2012). Public home care professionals' experiences of being involved in food distribution to home-living elderly people in Sweden – A qualitative study with an action research approach. *Journal of Nursing Education and Practice*, 2(2): 41–51. doi: 10.5430/jnep.v2n2p41

Reason, P. & Bradbury, H. (eds.) (2008). *The SAGE Handbook of Action Research: Participative Inquiry and Practice*. London: Sage.

Rojas, A., Black, J., Orrego, E., Chapman, G. & Valley, W. (2017). Insights from the Think&EatGreen@ School Project: How a community-based action research project contributed to healthy and sustainable school food systems in Vancouver. *Canadian Food Studies/La Revue canadienne des études sur l'alimentation*, 4(2): 25–46. doi: 10.15353/cfs-rcea.v4i2.225

Scrinis, G. (2013). *Nutritionism: The Science and Politics of Dietary Advice*. Columbia: Columbia University Press.

Soulard, C. T. & Lardon, S. (2019). Action-research helps Researchers foster smart rural development: Two case studies on local food policy. *Systemic Practice and Action Research*, 32(2): 155–166.

Torbert, W. R. & Taylor, S. S. (2008). Action inquiry: Interweaving multiple qualities of attention for timely action. In P. Reason & H. Bradbury (eds.), *The Sage Handbook of Action Research: Participative Inquiry and Practice* (pp. 239–251), London: Sage. doi: 10.4135/9781848607934.n24

Warrer, S. D. & Leer, J. (2018). Forhandlinger af mangfoldighed i danske kommunernes kostpolitikker for dagtilbud. *Nordisk Barnehageforskning*, 17(1): 1–18.

Williams, P. W. (2014). I would have never?: A critical examination of women's agency for food security through participatory action research. In J. Page-Reeves (ed.), *Off the Edge of the Table: Women Redefining the Experience of Food Insecurity* (pp. 275–313). Washington, DC: Lexington Books.

13

FOOD, DESIGN, AND DIGITAL MEDIA

Fabio Parasecoli

Visual digital media and material culture

How often do we check our social media accounts and see pictures of nicely plated dishes, elegantly set tables, and appealing restaurants and cafés? What elements make these images pleasant? To what extent do they influence our preferences, choices, and behaviours? Visual materials generate expectations about what food should look and be like, not only in itself but also in terms of places of production and consumption. The way things appear in pictures are at least partly expression of the design of the objects depicted in them: chefs are quite intentional in the way they present their culinary creations, arguing that customers eat with their eyes first; dishes and silverware are chosen to convey specific moods and styles, and are arranged with great care; the spaces surrounding food generate meaning through architecture, interior design, and lighting, as sensory environments influence our perceptions, our behaviours, and our individual and collective experiences from the point of view of affect (Wrigley & Ramsey, 2016).

The capacity to observe, assess, and analyse these design elements is increasingly important, as contemporary food trends tend to circulate through visual means, ranging from still pictures and moving GIFs to short ephemeral clips and longer, permanent videos (Calefato et al., 2016; Lupton, 2018). We are both recipients and producers of these visual materials, as we post and share our own. As this mode of communication is increasingly relevant in social interactions, it is imperative to be able to leverage the growing familiarity with digital media as a tool to gather and produce information not only on the use of media themselves, but also on how they influence and shape material culture.

This chapter focuses specifically on the use of digital media in the research about the material culture surrounding food, which includes practices, objects, and built environments that are often designed. By material culture, we refer to the aspects of social life that revolve around things and their materiality, including the materials they consist or are made of, how they are built (if human made), the functions they afford, and how users interact with them, both as individuals and members of communities (Buchli, 2002; Hicks & Beaudry, 2010; Tilley et al., 2006). In the case of food, material culture also refers at how people relate to the material qualities of what they eat through flavours, smells, sounds, textures, and other sensory characteristics. As we will see, digital technologies also need to be considered in their material aspects and functions, and not only in how they are used to represent food.

Through digital technologies and media, the material aspects of food culture are immediately integrated in the discursive dimension that always accompanies, defines, and at times critiques them in the communication process: how do people think of those things? How do they frame them? What ideas, values, and attitudes do they apply to their representations, consciously or unconsciously? Digital media such as websites, social media platforms (Yelp and Facebook, for instance), but above all visual apps such as Pinterest, Instagram, Snapchat, TikTok, and YouTube-like repositories, make traces of the material aspects of social life increasingly available and accessible online. By so doing, they also turn them into fodder for complex communication dynamics in which design often plays a major role as an element of appeal. In fact, digital media do not limit themselves to reflect contemporary realities of food manufacture, distribution, and consumption: they also constitute a major driver in generating, shaping, and supporting such trends. They disseminate the practices and values that underlie them while priming them for an unprecedented global circulation (Leer & Klitgaard Povlsen, 2016; Lewis, 2020; Rousseau, 2012). As they have turned into crucial primary sources – especially when examining the material aspects of evolving contemporary food trends – digital media can be used in food studies an effective and accessible research tools.

Besides material components of food culture, also the surrounding practices can be recorded and analysed through digital media. If the topic of our research is food at the zoo (as has been in one workshop I co-led in the Integrated Food Design program at Copenhagen University), it is relevant to take pictures not only of the food itself and its immediate containers, but also of how people interact with the food in the space where it is sold and consumed, including the architectural elements that surround it (from food stalls and kiosks to tables, benches to sit on, and garbage bins for waste disposal) and the spatial arrangement of the zoo. Where do people walk? Where do they rest? Where to they prefer to take a break and eat a snack? Where do they like to sit for a meal?

All these elements are not random, but carefully thought out, planned, and arranged: in other words, designed. Observing material culture and design in visual digital media can provide an entryway to explore different aspects of the way our societies deal with food. What kind of objects appear? How are they represented? Are they designed, arranged, or do they appear in a "candid shot" (or something that is supposed to look like one)? Who uses them in the images, and how? Do they confer value or status to users? How do the material qualities of objects generate affect in users and viewers? How do people feel about things? Are spaces constructed or "natural" (if such a category exists)? What does the camera linger on? How do images interpret the spaces? When it comes to interior design, what environments are represented? What emotions do lighting and shot framing convey? How do flows of people and objects move in them? How do people interact with space?

Very importantly in food studies, the production of such visual material is not limited to actors on the manufacturing side (chefs, artisans) or to those involved in communication and marketing (advertising agencies, photographers, and, more recently, influencers) but extends to consumers. We are all invited to post, add hashtags to our pictures, tag people and places, thus becoming co-producers in the global circulation of objects, discourses, and practices. Our own habits can be influenced by what we see on social media, while contributing to establish trends and viral phenomena. For that reason, it is important we become aware of our own role not only as observers, but also as co-producers of digital media.

This complexity requires an interdisciplinary approach. The research and pedagogical methods illustrated in this essay sit at the intersection of various fields. The emphasis on visual and sensory ethnography, as expressed for instance in Sarah Pink's work, contributes to highlight materials, their physical characteristics, and the affect they generate in those who interact with them. How can ethnographic research on sensory elements, perceptions, and embodied memories can be recorded, thus undergoing a process of mediation? (Pink, 2015b). How can video and photography support ethnographic research in the realms of materials and sensations, and how can visual material be organised and analysed in order to produce and share knowledge? (Pink, 2015b). These methodological concerns also extend to the research in the emerging field of digital ethnography, which draws attention to modes of embodiment and embeddedness of the virtual sphere in everyday life, as well as its impact on relationships, self-expression, and the construction of individual and social identities (Hjorth et al., 2017). However, in cultural and anthropological analysis these approaches often tend to concentrate on discursive elements and on the socio-cultural dynamics of the online space in itself, without sufficiently emphasising its connection with and influence on material culture or its designed aspects (Underberg & Zorn, 2013). This essay builds instead on digital ethnographies that explore not only experiences, relationships, and social worlds but also things and spaces (Pink, 2015a).

The methodological and theoretical perspectives explored in this essay also grow out of the expanding field of design anthropology, with its renewed interest in materiality and design used as tools not only for analysis but also for participative forms of research that tend to blur clear-cut distinctions between researchers and their subjects (Atzmon & Boradkar, 2017; Clarke, 2018; Gunn et al., 2013). While this essay will focus on qualitative research, we cannot forget that data scraping and metadata analysis can also provide a great amount of information, as marketers know well. Gathering data about consumers' preferences, their location, and their daily habits has become big business, although in many countries privacy laws are being passed (more or less successfully) to limit the most exploitative dimensions of the phenomenon.

To provide examples of how these theoretical and methodological reflections can be applied in practice, in this essay, I will assess two different uses of visual digital media in material culture analysis. The first example is based on a research project that has turned into a book, in which in collaboration with scholars and authors from around the world we try to assess a specific aesthetic regime and design style in food consumption environments that has become visible worldwide and that we have named "Global Brooklyn." The very idea for the project derives from the focus on materiality and design as relevant aspects in understanding emerging contemporary food trends and their complex interactions with and in the online space. Furthermore, the visual digital media we were examining to explore Global Brooklyn became also research tools in themselves, allowing the creation of a visual repository. Working with and through digital media allows for forms of research crowdsourcing that – although well tested in design - have not been widely employed in food studies yet. Input can come from the general public, colleagues, and students through the use of hashtags in social media and the sharing of images in repositories which now tend to live in the cloud.

The second example builds on this research experience and turns it into pedagogical tools to be used in graduate and undergraduate courses in food studies. How can we learn to pay greater attention to the material culture aspects of what we see in digital media? And how can we turn the same digital media we comfortably use for communication, self-expression, or just fun into research instruments? Research, teaching, and learning experiences are closely related, as the same theoretical and methodological approaches were used in all of them. The last section of the essay will focus specifically on the challenges of using visual digital media when exploring material culture.

Exploring Global Brooklyn

The relevance of digital media to understand material culture and design came to the fore while investigating, together with my research partner Mateusz Halawa from the Institute of Philosophy and Sociology of the Polish Academy of Sciences in Warsaw, the diffusion of what can be described as Global Brooklyn: a recurring, vaguely codified set of sensory elements, practices, and discourses which materialise in coffee places and restaurants in cities throughout the world (Halawa & Parasecoli, 2019).

Among the highly designed material culture elements that can be easily identified in these establishments, we can mention exposed pipes and wiring, visible building structures (from slabs of concrete to exposed bricks), stripped concrete floors, reclaimed wood and metal, refurbished industrial lighting, Edison bulbs, industrial or second-hand, mismatched furniture, blackboards with menus and other kinds of information written in chalk using fanciful fonts, and potted plants. Eating and drinking in Global Brooklyn are thoroughly designed experiences which often try to hide their artifice behind a performance of authenticity. Global Brooklyn relies heavily on a post-industrial look. Even in the "ready-made" spaces where we see little architectural intervention, it reveals an intense work of "designing the invisible" (Penin, 2018), including services, interactions, and experiences. Professional or not, designers are key figures in determining the material culture of Global Brooklyn and the affect it generates.

Although it emerged in other large metropolises such as San Francisco and Berlin, consumers and entrepreneurs around the world have tended to identify it with the New York borough, which is only one of many epicentres for this trend but has acquired particular visibility in the collective imaginary due to the centrality of US media and popular culture. Global Brooklyn's designed materialities and services of rugged genuineness in postindustrial settings both reflect and support a celebration of manual labor and craft; shifts in taste judgement towards more "natural" flavours; the reflexive, knowledge-intensive aspect of the practices and strategies of its actors; and appeals to an ethos of anti-corporatisation and authenticity, interpreted as revaluation of originality through the desire to stay true to oneself and one's ideals, both socially and professionally. In all these aspects, we have noticed that the designed characteristics of the objects and spaces generate forms of affect among those who interact with them.

Digital media have played a central role in moulding and disseminating the looks and feel of Global Brooklyn, as its users become co-producers and co-creators of value through their participation in social media, by posting and commenting. Global Brooklyn embodies the foodways of the network society. Spaces and objects are produced in response to the global circulation of images and then are in turn photographed, discussed, and disseminated. Global Brooklyn's designed material culture unfolds in digitally augmented spaces within deterritorialised networked communities of practice which share and learn from each other online and create new criteria of value. As environments and dishes are set up to attract attention in the crowded online mediascape, digital media afford insights in the centrality of design in determining such sensory experience, while raising questions about new forms of sociality built around algorithms that users cannot control.

While visual digital media have supported the global diffusion of Global Brooklyn and its material aspects, they have also been central in the development of our research on the topic, providing virtual access to far-flung locations and facilitating the recruitment of contributors for an edited volume that explores Global Brooklyn in places as diverse as Bogota, Copenhagen, and Cape Town. When possible, contributors are also gathering visual material using the hashtag #globalbrooklyn, making their research visible while it takes place.

We have used digital images in different ways in successive moments of the project. The inspiration for the project itself, although originating from our own travel experiences and fieldwork observations in Rome, Warsaw, Bologna, Rio de Janeiro, and Calicut, among other places, was confirmed by the examination of visual material that was available online. As restaurants, cafés, and stores use pictures of their interior design, objects, clients, and staff to entice patrons, the number of available images is staggering. Many photos are actually featured in discussions about hipsters and food, often tagged with ironic hashtags that reveal a critical attitude towards this particular food culture, which is often depicted as vapid and pretentious. However, the images that illustrated the main points in this criticism not only provide interesting details of material culture, from the suspenders and the beards of the baristas to the fancy lettering on blackboards, but also make us aware that in the research we need to engage with how these tangible elements affect the experience of all actors involved.

The contributors to our volume have been sending us the pictures they have taken in the locations they have been investigating. Although it is problematic to publish photos on academic books, due to cost of colour printing, we plan to launch an online gallery that readers of the book will be able to access free of charge. My research partner and I will be its curators, and we hope that readers will contribute with their own pictures, in order to establish a growing repository to facilitate the study of this phenomenon in particular and, more generally, of the relevance of materiality and design in food-related environments. As we give lectures and talks on Global Brooklyn in places as diverse as New York, Warsaw, and Shanghai, we ask those attending the events to participate in our effort. We are also looking into adding web coding to the repository that could identify pictures tagged as #globalbrooklyn on Instagram and other social media, in order to create a gallery that constantly refreshes itself as app users around the world upload their own photos.

Digital media as pedagogical tools

The number of pictures available on Instagram has increased because I have also used the same #globalbrooklyn hashtag in a course on Food and Culture in the Department of Nutrition and Food Studies at New York University, where I am a member of the faculty. In the course, we have explored different cultural aspects of food and foodways. The course had a methodological component in which we learned about and applied various ethnographic research methods on a theme that change every semester. I introduced #globalbrooklyn in the semester in which our ethnographic work focused on hipster food culture.

I invited students to conduct a visual ethnography on the topic by posting at least once a week on their Instagram accounts, using the hashtag as a way to create a shared repository of visual material. They also were required to post captions for their pictures in which they explained why they had taken that picture, what it meant in the framework of the visual ethnography, and what observable elements were particularly relevant.

In the same course, I also assigned a digital ethnography. While in the visual ethnography, the students were asked to generate information and data by producing their own visual material, in the digital ethnography, they instead analysed material that was already available on social media and other digital resources (Caliandro, 2017; Markham, 2017). As virtual environments are increasingly relevant as arenas in which food-related discourses are produced, practices are generated and circulated, and material objects are discussed and dissected, it is important to develop critical approaches to assess these sources. For the digital ethnography assignment, students had to pick a Global Brooklyn establishment (a café, a restaurant, a store, for instance) and explore its presence online on social media, websites, and videos, among other outlets. They were invited to focus not only on discourses and practices, but also on material and design aspects. I also introduced them to semiotics theories and methodologies in order to be better equipped for this kind of visual analysis.

Digital media can also be employed in creating visual moodboards, which I have used in food design workshops led with Mateusz Halawa and in an online course at New York University on food in urban environments. These kinds of moodboards are collections of images that can be put together to evoke affect or concepts that we have found are particularly important in our research. As the visual moodboards do not require text, they stimulate us to pay greater attention to visual elements, which are central when exploring design issues. We can access repositories of images online, download and modify the files that contain them, and create compositions using software like PowerPoint or Adobe. Moodboards can be shared online, or projected in class to engage in discussion about research questions, the experience of the research, and its results.

Besides practices and discursive elements, by paying attention to the visual we can focus on the materiality of food itself and its impact on the senses. By developing a sensibility for design elements and details through the use of digital media, we can hone our skills in observing feel, textures, smells, and sounds that constitute food perceptions. Furthermore, emphasising the role that objects and environments have in shaping individual and communal experiences, we can also be stimulated to reflect on how design shapes material culture.

Digital challenges

There are of course advantages and difficulties in using visual data from digital media to conduct research on material culture and design in food studies. When the process is shared, crowdsourced, or interactive, increasing its chances of success, we have only limited control on it. In our Global Brooklyn project, while some contributors are totally attuned to the visual aspects of material cultures, having worked in design, art history, or media, others are less used to consider, analyse, and include images in their work, while preferring to rely on text. We have left the participation in the visual component of the research project quite flexible and voluntary, in order to avoid causing stress for those who are not particular familiar with visual methodologies. We have tried to explain the relevance of images, but we did not insist too much, also because the visual material will not necessarily be part of the printed edited book coming out of the research project, but it is likely to be made available on a companion webpage.

In the case of a class assignment that relies on images either produced by the participants or found by them online, it is important to provide methodological directions, clear indications regarding the rationale and goals of the exercise, as well as the expectations in terms of its deliverables. Particular guidance may be necessary with those among us in the humanities and social sciences who are usually not fully attuned to observing and reflecting about the material and physical aspects of their research. They are generally even less used to look at design as an important component of such materiality, which connects it to cultural and productive aspects of the food system.

We may lack the visual literacy to actually interpret what we have photographed. At times, we focus on people and their interactions and completely ignore how human subjects interact with the designed environment and with objects. If we are not able to interpret objects and signs in order to assess the denotations and connotations they convey, we may need assistance in exploring how things generate meanings that are socially mediated and negotiated.

When working through digital media for the acquisition of visual material both for research and as a pedagogical tool, it is necessary to distinguish between literacy, the ability to perform scripts suggested by things according to common and accepted affordances, and competence, the capacity to "understand the conventions and functions of the respective genres." (Bernstein, 2009: 75). Somebody who has the literacy to operate a digital cam on their phones, may not necessarily have the competence to produce pictures that correspond to the "ethnographic research" genre and that can be successfully used to that aim. For generational reasons, older individuals may be less attuned to taking pictures on their smartphones or to sharing them on social media. In that case, they need further support to develop new literacies and competences. Others instead may have developed a professional habitus either as "influencers" or food bloggers, so they are likely to expect a very high level of aesthetic accomplishments for each picture, totally missing the point that immediacy and messiness can actually be an advantage when doing ethnographic research. Others may have had very negative – at times traumatic – experiences with social media, and bristle at the idea of using them to do research. In that case, it may be best to share pictures privately by email.

Working on food-related material culture and design through digital media can introduce us to methodologies and approaches that are not necessarily very common in food studies, increasing the crosspollination with the theories and methodologies of design, a field of theory and practice that has expressed growing concern and interest in what and how we produce, distribute, eat, and dispose of food.

In particular, reflections on design elements can provide entryways into research and theories about the agency of things and objects, that is to say their ability to have consequences and effects on food systems, foodways, and their human stakeholders, a concept that is quite central in food-related design and design anthropology (Parasecoli, 2018). Things shape human behaviour, as designers know very well. They evoke forms of embodied knowledge and performance that are not necessarily discursive. They train human bodies to move and locate themselves in space in particular ways, and to execute automatic gestures that can be quite outside of voluntary control. Marcel Mauss had already identified the "techniques of the body," that is everyday movements that are partially naturalised, remain largely unconscious, and very often are connected with or caused by interactions with objects (Mauss, 1973). When as children we learn how to eat, we may be taught to use forks, chopsticks, or our hands, to the point they become second nature and change the way we relate to food and the act of eating. We may be taught to eat while sitting on a chair, or sitting on the floor, or squatting. Each of these modalities is heavily depending on objects and built environments, whose design elements can be discussed through the use of visual material available in digital media.

Besides a reflection on the influence of material culture on our eating habits, it can be useful to consider the impact of the design of the very tool we use to take pictures (usually our smartphones) on the way we do research. In the case of these portable appliances, the normalised impulse is to constantly check them, to always have them with us, and to allow them to regulate our days, our calendar, and our activities. When using Instagram for research, we need to remind ourselves that we are using a specific technology that we access through a device that has its own materiality, determined by design strategies that favour certain affordances while discouraging others.

The smartphone has become a physical presence in our lives: we experience its weight in our pockets, its presence near us on tables and desks, and its reassurance in our hands. What kind of scripted actions and behaviours does it favour or discourage? How does its very designed materiality influence the way we take pictures, where and when we feel comfortable doing it, in which social contexts we perceive it as acceptable? A smartphone is a very different object from a camera, even a digital one, a specialised tool that immediately placed photographer, subjects, and environment in a socially recognisable relationship, which entails a certain formality and relies on a performance script that has precise cultural meaning.

The pictures taken with the smartphone can be more informal; they can also be taken stealthily, which poses ethical issues in terms of research. Do we need to get permission to take pictures (see section Ethical Considerations at the end of the chapter)? If we are in a public space and we take pictures of subjects in their public function we usually do not, but how about other people that end up in our shots? Also, what kinds of pictures are possible? That may depend very much from the smartphone we are using. Many brands have focused enormously on the quality of their cameras, used as a central feature to entice consumers to buy them. The most advanced ones allow for semi-professional pictures. Furthermore, high-resolution pictures are now quite current, as smartphones have increasingly larger processors and memories, and it is possible to keep one's pictures in the cloud, without cluttering one's devices. The facility of taking pictures makes it easier to gather a great number of them. When pictures were taken on film, and even later, when memory cards still had limited capacity, users were forced to be more cautious about the quantity of shots they would take. In the first case, films were expensive, and one would not see the results of one's research until the pictures were printed, a lengthy and expensive process. In the second case, the limited space available in terms of bits and bytes made researchers more aware of what they decided to keep, leading them to erase the pictures they thought were less valuable.

By facilitating a deeper evaluation of the impact of design and material culture on food experiences, visual research through digital media can also help us understand how material qualities of things (the so-called qualia) can provoke affective responses that can be coopted in social, political, and ideological dynamics (Fehérváry, 2013). Besides the meaning that design can impart on objects in terms of cultural references and mode of use, there are elements in their materiality itself that can influence moods and feelings. Imagine seeing a birthday cake that is covered in greyish cream cheese or a cheese that presents an extremely rough texture: most eaters would not perceived either as inviting, although they may be perfectly edible and not directly connected with any specific cultural negative values. Dealing with those foods would elicit specific affects that are as important as discursive or practice-based aspects (Harkness, 2015). Although thinking in these terms can be challenging if we are not attuned to reflecting about materiality, objects, and design, using these research methodologies in food studies research and pedagogy can open fruitful perspectives and generate multidisciplinary approaches, possibly facilitating collaborations with experts in different fields.

Dos and Don'ts

Make sure to assess your familiarity with the analysis of digital material in terms of material culture and design. If necessary, look for guidance and methodological tools.

If you use social media as a research tool, create fun and easy to remember hashtags to entice others to contribute.

Do not focus too much on discursive and textual data, in order to shift attention towards material culture and design.

Reflect on your own experience, and how the use of digital media for research differs from your everyday interactions with social media.

Do consider your own relationship with digital media as a standard: others may have very different experiences and approaches, and you need to take them into consideration.

Now you try

Choose a topic you can easily relate to: if you are intrigued by it or find it exciting, it is much more likely that you will participate actively in the research. Reflect on your own experiences and interactions with the objects and the physical spaces that surround food.

Create a hashtag that encapsulates the theme of the topic of the assignment and will generate a shared repository of images on social media.

Besides sharing images, you should also provide captions that explain why you have taken those pictures, how they fit in the theme of your research, what elements of material culture design they highlight, and how these influence food experiences.

Questions for reflection

How do material objects and built environments influence the embodied experiences of food production, distribution, and consumption for individuals and communities?

How does design shape the form, function, and meaning of material objects and built environments that have to do with food?

How do digital media shape our perception and appreciation of the design elements connected to food?

How do digital media contribute to the global circulation and success of specific food-related design elements?

How does visual research conducted through digital media contribute to the study of the role of design in food systems?

Ethical considerations

As digital media have become a central tool in visual research on design and the connections between design and food, ethical concerns need to be taken into consideration.

There are of course no problems regarding pictures of objects, built environments, interiors, and public spaces. However, objects are used by humans, who also inhabit designed places. It is generally acceptable to take pictures of individual in public spaces without their informed consent. In particular, it is acceptable to take pictures of individuals in their public functions (street vendors selling food, baristas making coffee, chefs cooking). Of course, issues of anonymity remain when it comes to consumers, so it is preferable to use pictures that convey information about the material objects and the built environments without focusing specifically on anybody. It is even better if people are photographed from behind, or from far enough that their features are not recognisable.

It is always advisable to verify with the offices in your institution that are in charge of ethical issues concerning research, variously named research ethics boards (REBs), institutional review boards (IRBs), ethical review boards (ERBs), and independent ethics committee (IECs). Rules and conventions may change from institution to institution, and from country to country. However, it is usually not necessary to obtain approval from such institutions if the pictures are only used for class assignments and not made public.

Further readings

Dürrschmidt, J. & Kautt, Y. (2019). *Globalized Eating Cultures: Mediation and Mediatization*. Cham: Palgrave Macmillan.

Horowitz, J. & Singley, P. (2004). *Eating Architecture*. Cambridge, MA: MIT Press.

Parasecoli, F. & Halawa, M. (2019). Rethinking the global table: Food design as future making. In M. Rosenthal & C. Flood (eds.), *Food: Bigger Than the Plate* (pp. 80–89). London: Victoria and Albert Museum.

Vodeb, O. 2017. *Food Democracy: Critical Lessons in Food, Communication, Design and Art*. Chicago, IL: Intellect.

References

Atzmon, L. & Boradkar, P. (eds.). (2017). *Encountering Things: Design and Theories of Things*. London: Bloomsbury.

Bernstein, R. (2009). Dances with things: Material culture and the performance of race. *Social Text*, 27(4): 67–94.

Buchli, V. (ed.). (2002). *The Material Culture Reader*. Oxford: Berg.

Calefato, P., La Fortuna, L., & Scelzi, R. (2016). Food-ography: Food and new media. *Semiotica*, 211: 371–388. doi: 10.1515/sem-2016-0087

Caliandro, A. (2017). Digital methods for ethnography: Analytical concepts for ethnographers exploring social media environments. *Journal of Contemporary Ethnography*, 47(5): 551–578.

Clarke, A. (ed.). (2018). *Design Anthropology: Object Cultures in Transition*. London: Bloomsbury.

Fehérváry, K. (2013). *Politics in Color and Concrete: Socialist Materialities and the Middle Class in Hungary. New Anthropologies of Europe*. Bloomington: Indiana University Press.

Gunn, W., Otto, T., & Smith, R.C. (eds.). (2013). *Design Anthropology: Theory and Practice*. London: Bloomsbury.

Halawa, M. & Parasecoli, F. (2019). Eating and drinking in Global Brooklyn. *Food Culture and Society*, 22(4): 387–406. doi: 10.1080/15528014.2019.1620587

Harkness, N. (2015). The pragmatics of qualia in practice. *Annual Review of Anthropology*, 44: 573–589. doi: 10.1146/annurev-anthro-102313-030032

Hicks, D. & Beaudry, M. (eds.). (2010). *The Oxford Handbook of Material Culture Studies*. Oxford: Oxford University Press.

Hjorth, L., Horst, H., Galloway, A., & Bell, G. (eds.). (2017). *The Routledge Companion to Digital Ethnography*. London: Routledge.

Leer, J. & Klitgaard Povlsen, K. (eds.). (2016). *Food and Media: Practices, Distinctions and Heterotopias*. London: Routledge.

Lewis, T. (2020). *Digital Food: From Paddock to Platform*. London: Bloomsbury.

Lupton, D. (2018). Cooking, eating, uploading: Digital food cultures. In K. Lebesco & P. Naccarato (eds.), *The Bloomsbury Handbook of Food and Popular Culture* (pp. 66–79). London: Bloomsbury.

Markham, A. (2017). Ethnography in the digital internet era. In N. Denzin & Y. Lincoln (eds.), *Sage Handbook of Qualitative Research* (pp. 650–668). Thousand Oaks, CA: Sage.

Mauss, M. (1973). Techniques of the body. *Economy and Society*, 2(1): 70–85.

Parasecoli, F. (2018). Food, design, innovation: From professional specialization to citizens' involvement. In K. Lebesco & P. Naccarato (eds.), *Handbook of Food and Popular Culture* (pp. 27–39). London: Bloomsbury.

Penin, L. (2018). *An Introduction to Service Design: Designing the Invisible*. London: Bloomsbury Visual Arts.

Pink, S. (2015a). *Digital ethnography: Principles and Practices*. London: Sage.

Pink, S. (2015b). *Doing Sensory Ethnography*. London: Sage

Rousseau, S. (2012). *Food and Social Media: You Are What You Tweet*. Lanham, MD: Altamira.

Tilley, C., W. Keane, S. Kuchler, M. Rowlands, & P. Spyer (eds.). (2006). *Handbook of Material Culture*. London: Sage.

Underberg, N. & Zorn, E. (eds.). (2013). *Digital Ethnography: Anthropology, Narrative, and New Media*. Austin. University of Texas Press.

Wrigley, C. & R. Ramsey. 2016. Emotional food design: From designing food products to designing food systems. *International Journal of Food Design*, 1(1): 11–28. doi:10.1386/ijfd.1.1.11_1.

PART 4

Digital archives and network analysis

14

ARCHIVED WEB AS A RESOURCE IN FOOD HISTORY

Caroline Nyvang

Today, the World Wide Web is the platform for a number of cultural phenomena that previously used to leave their marks offline. Emails have almost supplanted the physical letter, paper diaries have been swapped for online blogs, and we befriend each other virtually via social media networks. This turn towards the digital also includes how we create and consume food and recipes, as food blogging, like other virtual outlets, has "emerged as a new and viable way for people to share information about food in a non-professional capacity" (Lofgren, 2013, p. 1). However, online expressions are fleeting and need to be stored before they lend themselves to being studied. Focusing on methodological possibilities and challenges from different examples of use, this chapter explores how various Internet archives, like the archived web in general, can be fruitfully employed in food studies. The chapter is based on the premise that paying attention to archiving strategies is central to understanding the potential of a given source material. Accordingly, it focuses on how web archives differ from traditional physical archives.

Sharing food in the digital age

In a number of ways, what we eat seems especially geared to the digital turn. As had been shown by a number of scholars long before the widespread use of the Internet, sharing food is a core aspect of the unique human–food experience (Bloch, 1999; Fischler, 1988; Mennell et al., 1992; Rozin, 1999). Yet a meal is also inherently ephemeral, often disappearing from our plates faster than it took to concoct. In the so-called selfie culture, in which we invite millions of onlookers into our quasi-private sphere, displaying what we had for dinner serves as a longer lasting bouillon cube of who we are and the values we hold. This is surely part of the explanation why, on the photo- and video-sharing service Instagram, food images are one of the most popular subjects, exceeded only by the selfie (Amato et al., 2017, p. 1333).

Furthermore, the recipe, that familiar format through which we have transferred culinary knowledge to both our relatives and the broader public, is especially in tune with the ways in which information is generally shared and used online. Sometimes called the "co-creative collective" and at other times scoffed at as a cut-and-paste culture, the digital age allows a hitherto unseen mosaic of appropriation. In her seminal work on the recipe, Susan Leonardi argues that "a recipe is reproducible, and, further, its hearers-readers-receivers are *encouraged* to reproduce it and, in reproducing it, to revise it and make it their own" (Leonardi, 1989, p. 344). Others have followed Leonardi's lead, claiming that the recipe is fundamentally "unauthorized," as it is rooted in a continual process of "combing, changing and adding to old recipes" (McDougall, 1997, pp. 107, 117). This might help us understand why, at the time of writing, recipes and culinary creations are still not granted the same copyright protection as other literary and artistic works (Janssens, 2013; Lawrence, 2011): they are usually considered the product of a collective and historical effort.

Although the web is famed for allowing the widespread distribution of information, it is not simply a matter of *more* of the same. As emphasised in 1964 by Marshall McLuhan, whose work has become a keystone in modern media theory, a medium is far from being an empty container (McLuhan, 1964). The characteristics of different media inevitably affect the message, and the web too helps shape and form the way we approach food not only online but also offline. A number of studies have convincingly shown, and theorised, how the World Wide Web has changed the dynamics of restaurant reviews (Kobez, 2018; Rousseau, 2012), the potential for food activism (Schneider et al., 2017; Schneider & Davis, 2010), food marketing (Freeman et al., 2014; Montgomery & Chester, 2009), etc.

The ephemeral web

All these studies outline interesting food dynamics fostered by the Internet, and they open up exciting new avenues for further exploration. However, online information is not as perennial as many would like to think. According to the Library of Congress, the average lifespan of a webpage is around a hundred days, and most of the lost content is no longer retrievable – anywhere (SalahEldeen & Nelson, 2012; Tringham & Ashley, 2015, p. 33). The loss of information seems to be especially pronounced when it comes to dynamic content, such as blogs, social media posts, etc., which to a high degree are user-generated (Barone et al., 2015). In short, in a matter of years, all content that is not preserved in real time will likely be altered or gone for good.

In conclusion, we need to archive the web if we wish to study it. For the past decade, a number of commercial actors have offered the preservation of web materials by means of online services or locally installed software that enables users to establish private repositories of particular web components, specific websites, and even live on-screen action. These kinds of services allow for what is sometimes termed "micro-archiving" (Brügger, 2011, p. 25) or "small scale Web archiving" (Lecher, 2006). While these might help us preserve web materials for later studies, the usability of the content still largely hinges on whether we managed to capture what we needed in real time. Furthermore, these private and often locally stored archives seldom allow us to give others access to our data, nor do they enable us to create reliable and correct references to the materials on which we base our studies (Nyvang et al., 2017). Thus, if one wishes to meet the most common standards of good scientific practice, microarchives will not suffice.

Macroarchiving, carried out "in order to archive ... cultural heritage in general" (Brügger, 2011, p. 25), is an intricate task which, at the time of writing, is best carried out by dedicated public institutions with the proper legal authority and the financial and technical means to do so. Since the mid-1990s, a number of educational establishments, libraries, and private companies have systematically been archiving materials published online. A survey conducted in 2014 identified more than forty different initiatives, preserving a total of 6.6 petabytes of data (Costa et al., 2016), or the equivalent of 500.000.000.000 (five hundred trillion) pages of standard typed text. One can safely assume that this number has greatly increased since then, although web archives do not tend to grow at the same impressive rate as the Internet itself.

The first web archives were established at the dawn of the World Wide Web. One of the pioneers was Internet Archive, an American non-profit company, which since 1996 has aimed to archive the publicly available Internet in its entirety. In reality, this proved to be a "Sisyphean improbability" (Rogers, 2018, p. 43), as the Internet is ever expanding, though the Internet Archive is still by far the most comprehensive web archive. The lion's share of the archive has been collected by means of so-called "crawlers" that systematically visit and harvest websites. From 2001, this material has been searchable through an interface called the Wayback Machine, an easy-to-use search engine that virtually transports you back in time.

Although the Internet Archive is often touted as "the most complete archive of webpages in the world" (Tranos & Stich, 2020, p. 2), smaller initiatives that frequently capture distinct parts of the web will likely provide better coverage of national domains and distinct topics. But while the Internet Archive is publicly accessible and open to all, access to many local initiatives is often very limited for reasons of copyright or other legal restrictions (Brügger & Schroeder, 2017, p. 10).

Since the turn of the millennium, libraries and archives have started to archive websites in accordance with national legal deposit legislation, which entitles institutions to collect e-materials. As opposed to the – at least in principle – all-encompassing Internet Archive, web archives such as the Danish Netarkivet, the Finnish Web Archive and the UK Web Archive all focus on national domains. These initiatives often supplement broad domain crawls with selective efforts in order to create special collections that secure in-depth coverage of unforeseen as well as foreseen events such as disasters, social protest movements, and elections (Rogers, 2018; Rollason-Cass & Reed, 2015).

Alongside these national archival endeavours, a number of organisations have taken on the task of creating web archives dedicated to particular topics. One example is the Human Rights Watch Archive, based at Columbia University (Webster,

2019, p. 3). Another is The End of Term Web Archive, which aims to capture and save US government documents at the end of each presidential term.

Approaching archived web

As human beings, we have had more than a 1,000 years of experience with physical archives and the corresponding archival systems, but digital archives in general and web archives in particular differ significantly from their analogue counterparts (Cook, 2013; Theimer, 2011). While some of the pertinent challenges might be eased in the course of time, a number of these concern the very nature of the archived web and must prompt methodological discussions and development.

The traditional paper archive is usually defined with respect to provenance, unified collections, and their being kept in their original order or context (Theimer, 2012). As a fourth significant feature, Thiemer adds that traditional archives primarily consist of "original or unique materials and not published ones" (Theimer, 2012). The latter has prompted Brügger to argue that "web archive" is a misleading term, as archived web collections are more comparable to library holdings of published materials (Brügger, 2017). Hence, Brügger suggests that it would be more consistent to use the term "webraries" to denote collections of web materials that have previously been made available to the public, whereas "web archive" would be the suitable term for collections of the minutes of private meetings, a company's Intranet, etc. However, Brügger (2017) also encourages us "to use the verb 'Web archiving' for the activity of collecting and preserving the Web, regardless of which part of the Web is in question" (p. 187).

When it comes to digitally created material, the very principles that define an archive are being contested in other ways too, posing a number of challenges to researchers. First and foremost, as web archives primarily consist of unappraised data that have been harvested automatically, they typically contain very large quantities of unstructured and messy data, that is, they are not arranged in unified collections that readily lend themselves to studies of a particular cultural phenomenon. Furthermore, many of the publicly available web archives – most notably the Internet Archive – do not feature free text search. Instead, most web archives rely on the Wayback Machine, aptly named after the WABAC time-travelling machine from the American animated series *The Rocky and Bullwinkle Show*. And just like its cartoonish predecessor, the Wayback Machine's user interface transports its users back in time, enabling one to revisit past websites via the original URL. This means that one has to know the exact domain name of a website or be utterly creative to access historical websites.

Nevertheless, the difference between the archived web and traditional archives also holds out a promising potential. The vast amounts of data involved have allowed researchers to explore histories of distinct online phenomena such as the dawn of blogging, the heydays of online gaming and the demise of musical piracy (Brügger & Milligan, 2018). However, the potential of the archived web has yet to be fully utilised in food studies.

Further explorations

As all of the above implies, the number of food studies that have made use of the archived web is close to zero. This is most likely due to the fact that there are not yet any established methodologies to help scholars access the different internet archives (Brügger, 2018). While there have been efforts to develop some methodologies, such as Virtual Ethnography (Hine, 2000), Computer-assisted Webnography (Horster & Gottschalk, 2012), and Netnography (Kozinets, 2013), these are all built around participant observation. This seems to be entirely suitable for studying the live web, but does not provide any tools for handling archived materials. Furthermore, there are no dedicated web archives that offer neatly labelled boxes containing recipes, food blogs, pictures of meals, etc. The data and delimitations will have to be dug out of terabytes of other cultural expressions.

However, web archives do offer unique opportunities for analysing representations of food in the context of media history, thus adding new dimensions to our understanding of current food practices. The archived web will enable us to further scrutinise particular online phenomena that have already been given scholarly attention, such as the new possibilities for health-monitoring afforded by the Internet (Kent, 2018) and online communities centred around different dietary regimes (Boero & Pascoe, 2012; Dinhopl et al., 2015; Pirkey, 2015). It will also enable us to do this in a historical context, as well-known cultural practices both shape and are shaped by the advent of new media technology. One example that springs to mind is the transmedia history of the recipe, but this is far from being the only topic that might be fruitfully explored by means of the archived web. As we increasingly create and consume news in transient digital formats, archived newssites are crucial materials if we wish to be able study how food is mediated. The twenty-first-century restaurant industry is another topic that would surely benefit from being studied through the inclusion of the archived web. Restaurant reviews are now mostly published online, where guests and chefs are also able to say their piece. Furthermore, restaurants are now almost exclusively promoted via webpages and online advertisements.

The way the archived web is usually organised and made accessible – that is, as distinct URLs – make its collections especially suitable for historiographies of single webpages. This approach has previously yielded interesting results. For instance, by conducting a morphological and syntactical analysis of the web elements on *The Guardian* website from 1996 to 2015, Bødker and Brügger (2017) investigated the ways in which journalism has been affected by the possibilities of the digital.

In a similar vein, consulting archived webpages from the last 10 years can be used to investigate how the online recipe repository Allrecipes.com has changed in its content, design, features, and language. The archived web can also be employed to make sense of how the various possibilities afforded by online media have impacted the commercialisation strategies of celebrity chefs such as Nigella Lawson and Jamie Oliver.

The archived web has also proved useful in the analysis of quantitative texts. For instance, by revisiting Australian webpages from 2005 to 2015, Ackland and Evans documented significant changes in the language used by both sides in the abortion debate (2017). A similar approach could be used to detect linguistic changes in relation to food.

For those wishing to embark on a journey into the archived web, there are a number of highly relevant methodological pointers. First and foremost, it is important to remember that the process of archiving the web fundamentally alters the material. As opposed to physical documents that are kept in a conventional archive, the representation of a particular website will never be exactly identical to that which was originally stored (Nielsen, 2016, p. 8). This means that one needs to pay close attention to each web archive's individual settings. Furthermore, as no archive is able to mirror reality, whether it stores papers or web materials, consideration of each web repository's harvesting strategy is crucial in order to be able to describe the limitations of one's research. The individual archive usually provides relevant information about scope and settings, although it sometimes requires a bit of investigation.

The archived web can be employed using a variety of established methods. A number of examples have a promising potential that could also add great value to digital food studies. For instance, the great number of captured websites has allowed quantitative studies such as historical link analyses, which explore how different websites have been connected (Brügger, 2013; Hale et al., 2014), and text-mining that bares textual patterns across a large number of websites (Yeung & Jatowt, 2011).

Furthermore, the archived web also lends itself to qualitative studies of the historiography of single sites (Brügger, 2009; Foot & Schneider, 2010), as well as of the language and imagery of particular websites (Gorsky, 2015; Nyvang & Vallgårda, 2018). In fact, most of the research themes discussed in this book could be explored in a contemporary historical setting through the archived web. The relevant topics will be decided by the scholars of the future, but one thing seems obvious to us all: it is hard to imagine a study of the twenty-first century that does not include the World Wide Web.

Dos and Don'ts

Do pay attention to how different web archives differ from one another, in terms of both coverage and accessibility. It is sound methodology to describe the limitations of the particular archives you have used in order for others to be able to understand the reach of your analysis and conclusions.

Don't ever assume that your findings in a web archive are based on a complete data set. The web is continually expanding, and no repository has yet managed to archive everything. Bear in mind that, although the user interface is based on familiar technologies and designs, such as the Google search engine, a repository in the archived web is *not* comparable to the live web.

Whenever possible, try to make references to the archived web instead of to live web pages. The web is in constant flux, and the text or images that you analyse may soon be altered or disappear altogether. This means that, although it conflicts with good scientific practice, other scholars will not be able to challenge your conclusions or reproduce your results.

Now you try

Using the publicly available Internet Archive [https://archive.org/], explore snapshots of different websites related to food. Among the useful sites you can explore are online media sites such http://cooking.nytimes.com/, online recipe collections, such as https://www.epicurious.com, and restaurant websites.

Try tracking a particular food site through time. How has the website evolved through time? What kinds of functionality have been added or removed? Have the layout, the colour scheme, the font or the rhetoric changed? And if so, to what extent do these changes reflect larger changes in our approach to food?

Questions for reflection

Does the way food is mediated online differ significantly from how analogue sources, such as handwritten recipes, cookbooks, and food magazines, have portrayed food? If so, how? What has prompted these changes?

How can digital media affect the way we handle and perceive food?

Ethical considerations

While there is yet no one codex that clearly defines your ethical obligations when researching and relaying online information, it is considered best practice always to balance the individual's right to privacy with the researcher's right "to pursue knowledge for the benefit of society" (Lomborg, 2018, p. 101).

The archived web contains a hitherto unseen amount of personal and sensitive information that can easily be made searchable. And although what is stored has usually been willingly published online in a semi-public domain, this often happens unbeknownst to the individuals who have shared the information, and it might even include material that might have been deleted at one point.

As a rule of thumb, it is always beneficial to reflect on what you decide to disclose in your research and to consider this against the vulnerability of the subject or community that you are studying (Markham & Buchanan, 2012, p. 4)

It is good practice to anonymise text and images whenever this does not interfere with the major analytical points of your work. Also, remember that, while online expressions can be quoted under fair use or similar principles, you must always make sure to credit an author.

Further readings

Brügger, N. (2011). Web archiving: Between past, present, and future. In M. Consalvo & C. Ess (eds.), *The Handbook of Internet Studies* (pp. 24–42). John Wiley & Sons.

Brügger, N., & Milligan, I. (2018). *The SAGE Handbook of Web History*. London: Sage.
Markham, A., & Buchanan, E. (2012). *Recommendations from the AoIR Ethics Working Committee (Version 2.0)*. 19.

References

Ackland, R., & Evans, A. (2017). Using the web to examine the evolution of the abortion debate in Australia, 205–15. In N. Brügger & R. Schroeder (eds.), *The Web as History* (pp. 159–189). UCL Press.

Amato, G., Bolettieri, P., Monteiro de Lira, V., Muntean, C. I., Perego, R., & Renso, C. (2017). *Social media image recognition for food trend analysis. Proceedings of the 40th International ACM SIGIR Conference on Research and Development in Information Retrieval SIGIR '17*, pp. 1333–1336. doi: 10.1145/3077136.3084142

Barone, F., Zeitlyn, D., & Mayer-Schönberger, V. (2015). Learning from failure: The case of the disappearing web site. *First Monday*, 20(5). doi: 10.5210/fm.v20i5.5852

Bloch, M. (1999). Commensality and poisoning. *Social Research; New York*, 66(1), 133–149.

Bødker, H., & Brügger, N. (2017). The shifting temporalities of online news: *The Guardian*'s website from 1996 to 2015. *Journalism*, 19(1), 56–74.

Boero, N., & Pascoe, C. J. (2012). Pro-anorexia communities and online interaction: Bringing the pro-ana body online. *Body & Society*, 18(2), 27–57. doi: 10.1177/1357034X12440827

Brügger, N. (2009). Website history and the website as an object of study. *New Media & Society*, 11(1–2), 115–132.

Brügger, N. (2011). Web archiving:— Between past, present, and future. In M. Consalvo & C. Ess (eds.), *The Handbook of Internet Studies* (pp. 24–42). Oxford: John Wiley & Sons.

Brügger, N. (2013). Historical network analysis of the web. *Social Science Computer Review*, 31(3), 306–321. doi: 10.1177/0894439312454267

Brügger, N. (2017). Webraries and web archives: The web between public and private. In E. D. Baker & W. Evans (eds.), *The End of Wisdom?* (pp. 185–190). Oxford: Elsevier. doi: 10.1016/B978-0-08-100142-4.00023-3

Brügger, N. (2018). *The Archived Web: Doing History in the Digital Age*. Cambridge, MA: The MIT Press.

Brügger, N., & Milligan, I. (2018). *The SAGE Handbook of Web History*. London: Sage Publications. http://ebookcentral.proquest.com/lib/kbdk/detail.action?docID=5601770

Brügger, N., & Schroeder, R. (2017). *The Web as History: Using Web Archives to Understand the Past and the Present*. UCL Press.

Cook, T. (2013). Evidence, memory, identity, and community: four shifting archival paradigms. *Archival Science*, 13(2–3), 95–120. doi: 10.1007/s10502-012-9180-7

Costa, M., Gomes, D., & Silva, M. J. (2016). The evolution of web archiving. *International Journal on Digital Libraries.*. doi: 10.1007/s00799-016-0171-9

Dinhopl, A., Gretzel, U., & Whelan, A. (2015). Labeling as a Social Practice in Online Consumption Communities. *Psychology & Marketing*, 32(3), 240–249. doi: 10.1002/mar.20777

Fischler, C. (1988). Food, self and identity. *Social Science Information*, 27(2), 275–292.

Foot, K., & Schneider, S. (2010). Object-oriented web historiography. In N. Brügger (transl.), *Web History* (pp. 61–79). New York: Peter Lang.

Freeman, B., Kelly, B., Baur, L., Chapman, K., Chapman, S., Gill, T., & King, L. (2014). Digital junk: Food and beverage marketing on facebook. *American Journal of Public Health*, 104(12), e56–e64. doi: 10.2105/AJPH.2014.302167

Gorsky, M. (2015). Into the dark domain: The UK web archive as a source for the contemporary history of public health. *Social History of Medicine*, 28(3), 596–616. doi: 10.1093/shm/hkv028

Hale, S. A., Yasseri, T., Cowls, J., Meyer, E. T., Schroeder, R., & Margetts, H. (2014). *Mapping the UK webspace: fifteen years of british universities on the web.* In *Proceedings of the 2014 ACM Conference on Web Science – WebSci '14*, pp. 62–70. doi: 10.1145/2615569.2615691

Hine, C. (2000). *Virtual ethnography*. London: Sage.

Horster, E., & Gottschalk, C. (2012). Computer-assisted webnography: A new approach to online reputation management in tourism. *Journal of Vacation Marketing*, 18(3), 229–238. doi: 10.1177/1356766712449369

Janssens, M.-C. (2013). Copyright for culinary creations: A seven course tasting menu with accompanying wines. *SSRN Electronic Journal*. doi: 10.2139/ssrn.2538116

Kent, R. (2018). Social media and self-tracking: representing the 'Health Self'. In B. Ajana (ed.), *Self-Tracking: Empirical and Philosophical Investigations* (pp. 61–76). Springer International Publishing. doi: 10.1007/978-3-319-65379-2_5

Kobez, M. (2018). 'Restaurant reviews aren't what they used to be': digital disruption and the transformation of the role of the food critic. *Communication Research and Practice*, 4(3), 261–276. doi: 10.1080/22041451.2018.1476797

Kozinets, R.V. (2013). *Netnography: Doing Ethnographic Research Online*. London: Sage.

Lawrence, M. G. (2011). Edible plagiarism: Reconsidering recipe copyright in the digital age note. *Vanderbilt Journal of Entertainment and Technology Law*, 1, 187–224.

Lecher, H. E. (2006). Small scale academic web archiving: DACHS. In J. Masanès (ed.), *Web Archiving* (pp. 213–225). Berlin Heidelberg: Springer. doi: 10.1007/978-3-540-46332-0_10

Leonardi, S. J. (1989). Recipes for reading: Summer pasta, lobster à la riseholme, and key lime pie. *PMLA*, 104(3), 340–347.

Lofgren, J. (2013). Food Blogging and Food-related Media Convergence. *M/C Journal*, 16(3). http://www.journal.media-culture.org.au/index.php/mcjournal/article/view/638

Lomborg, S. (2018). Ethical considerations for web archives and web history research. In N. Brügger & I. Milligan (eds.), *The SAGE Handbook of Web History* (pp. 99–111). London: Sage Publications. http://ebookcentral.proquest.com/lib/kbdk/detail.action?docID=5601770.

Markham, A., & Buchanan, E. (2012). *Recommendations from the AoIR Ethics Working Committee (Version 2.0)*. 19.

McDougall, E. J. (1997). Voices, stories, and recipes in selected Canadian community cookbooks. In A. Bower (ed.), *Recipes for reading: community cookbooks, stories, histories* (pp. 105–117). Amherst: University of Massachusetts Press.

McLuhan, M. (1964). *Understanding media, the extensions of man* (2nd ed.). London: Sphere Books.

Mennell, S., Murcott, A., & Otterloo, A. H. van. (1992). *The Sociology of Food: Eating, Diet, and Culture*. London: Sage.

Montgomery, K. C., & Chester, J. (2009). Interactive Food and Beverage Marketing: Targeting Adolescents in the Digital Age. *Journal of Adolescent Health*, 45(3), Supplement, S18–S29. doi: 10.1016/j.jadohealth.2009.04.006

Nielsen, J. (2016). *Using Web Archives In Research: An Introduction*. NetLab. http://www.netlab.dk/wp-content/uploads/2016/10/Nielsen_Using_Web_Archives_in_Research.pdf

Nyvang, C., Kromann, T. H., & Zierau, E. (2017). Capturing the web at large: A critique of current web referencing practices. *Researchers, Practitioners and Their Use of the Archived Web*, 1–9. doi: 10.14296/resaw.0004

Nyvang, C., & Vallgårda, K. (2018). Sorg som opposition. *Temp - tidsskrift for historie*, 8(16), 59–80.
Pirkey, M. F. (2015). People like me: Shared belief, false consensus, and the experience of community. *Qualitative Sociology*, 38(2), 139–164. doi: 10.1007/s11133-015-9303-6
Rogers, R. (2018). Periodizing web archiving: Biographical, event-based, national and autobiographical traditions. In N. Brügger & I. Milligan (eds.), *The SAGE Handbook of Web History* (pp. 42–56). London: Sage.
Rollason-Cass, S., & Reed, S. (2015). Living movements, Living archives: Selecting and archiving web content during times of social unrest. *New Review of Information Networking*, 20(1–2), 241–247. doi: 10.1080/13614576.2015.1114839
Rousseau, S. (2012). *Food and Social Media: You Are What You Tweet*. AltaMira Press. http://ebookcentral.proquest.com/lib/kbdk/detail.action?docID=950441
Rozin, P. (1999). Food is fundamental, fun, frightening, and far-reaching. *Social Research*, 66(1), 9.
SalahEldeen, H. M., & Nelson, M. L. (2012). *Losing My Revolution: How Many Resources Shared on Social Media Have Been Lost?* http://arxiv.org/abs/1209.3026
Schneider, T., & Davis, T. (2010). Advertising food in Australia: between antinomies and gastro-anomy. *Consumption Markets & Culture*, 13(1), 31–41. doi: 10.1080/10253860903346740
Schneider, T., Eli, K., Dolan, C., & Ulijaszek, S. (2017). *Digital Food Activism*. Routledge.
Theimer, K. (2011). What is the meaning of archives 2.0? *The American Archivist*, 74(1), 58–68. doi: 10.17723/aarc.74.1.h7tn4m4027407666
Theimer, K. (2012, June 26). Archives in context and as context. *Journal of Digital Humanities*. http://journalofdigitalhumanities.org/1-2/archives-in-context-and-as-context-by-kate-theimer/.
Tranos, E., & Stich, C. (2020). Individual internet usage and the availability of online content of local interest: A multilevel approach. *Computers, Environment and Urban Systems*, 79. doi: 10.1016/j.compenvurbsys.2019.101371
Tringham, R., & Ashley, M. (2015). Becoming archaeological. *Journal of Contemporary Archaeology*, 2(1), 29–41. doi: 10.1558/jca.v2i1.27089
Webster, P. (2019). Existing web archives. In N. Brügger & I. Milligan (eds.), *The SAGE Handbook of Web History* (pp. 30–41). London: Sage Publications Ltd. doi: 10.4135/9781526470546.n3
Yeung, C. A., & Jatowt, A. (2011). Studying how the past is remembered: Towards computational history through large scale text mining. In Proceedings of the 20th ACM Conference on Information and Knowledge Management, 2011, p. 10.

15

TRACING CULINARY DISCOURSE ON FACEBOOK

A digital methods approach

Anders Kristian Munk

> "Alain Ducasse says you can stop your Dry January and go back to enjoying wine!"
>
> (FineDiningLovers on Facebook, January 17, 2020)

> "Anthony Bourdain's Final Book Will Be Released This Year."
>
> (FineDiningLovers on Facebook, January 17, 2020)

> "The Paul Bocuse Restaurant Has Lost its Third Star...are Michelin correct?"
>
> (FineDiningLovers on Facebook, January 17, 2020)

If you follow a page like FineDiningLovers on Facebook, you will be familiar with the feeling that names of certain chefs and food personalities always appear in the same places. It is, for example, no surprise to find Alain Ducasse and Paul Bocuse mentioned by the same food blog. Similarly, when a restaurant writes a post that acknowledges the achievements of a chef like Magnus Nilsson, one could reasonably expect this restaurant to also write posts about Massimo Bottura, Joan Roca or some other superstar from the San Pellegrino best restaurants list. When a blogger mentions Rick Stein, other British TV-chefs like Keith Floyd or Marco Pierre White could likely be next. Spheres, foams, and agar pearls rhyme with Ferran Adriá, Heston Blumenthal and molecular gastronomy. Distinct ways of talking about food come with distinct gastronomic reference points (chefs, ingredients, cooking techniques, terroirs, restaurants, you name it).

In this chapter, I will show you how to map such discursive patterns with digital methods. In a data set of 102M posts collected from 242K food-related Facebook pages worldwide, I track mentions of almost 700 chefs. I construct a "co-chef" network (a network of chefs connected to each other if they are mentioned by the same pages) which I subject to a visual network analysis. I provide the network file so that you can follow my steps but it is of course also possible to take a different direction on your own (e.g., based on a smaller data set, differently curated, or from a different medium). I also provide a series of intermediary visualisations annotated with my field notes which will hopefully make my thought process clearer.

I am interested in chefs, and Nordic chefs in particular, an interest I have been cultivating as part of a wider effort to understand how the New Nordic Food movement has developed and diffused over time (Munk & Ellern, 2015) and the multiple enactments of *terroir* more generally (Høyrup & Munk, 2007). I have previously mapped that development by following different ingredients, concepts, and practices (e.g., sea buckthorn or fermentation) in a multisited ethnographic study across Scandinavia (in the style of Marcus, 1995). Social media makes it possible to associate discussions about food with specific actors, in specific places, at specific points in time, which has opened up new opportunities for this kind of mapmaking (Munk & Jensen, 2015). Digital traces make it possible to follow how ingredients and cooking techniques are being adopted into culinary practices as these are exhibited and performed in text and images on the Web.

Chefs and other food personalities are, in this respect, a relatively straightforward trace to follow. When the same food blogger cites René Redzepi or Claus Meyer, it is typically a sign of recognition and we can be fairly certain that they are also referencing the New Nordic movement. Other traces are comparatively more challenging. We can, for example, have a presumption that sea buckthorn is an indicator ingredient for the New Nordic Cuisine, but there could be other sources of inspiration that would cause a food blogger to post a recipe with this particular ingredient.

The simplest approach to understanding how Nordic chefs have become household names outside of Scandinavia would probably be to search for mentions of them by pages outside the region. However, as stated in the opening paragraph, I already have a clear impression from casually browsing these pages that certain chefs tend to be mentioned together in distinct discursive patterns. So, rather than presuming in the research design that geography is what matters here I am going to adopt what Richard Rogers calls a "post-demographic" approach to social media (Rogers, 2009a) and assume that food talk on Facebook is organised around communities of interest rather than geography or language. Instead of asking where in the world Nordic chefs are talked about, I ask to what extent Nordic chefs have broken out of their local Scandinavian food-talk bubbles and migrated into other and more international discourses.

The difference is central to understanding the contribution of network analysis to this kind of food study. If we presume that discourses about food are national or regional, in the sense that there is a Danish conversation about food on Facebook which is qualitatively different from the French or the American conversation about food, then there is no need for networks. I know where the food-related pages in my data set are located geographically, what remains is simply to count how much Nordic chefs are mentioned by each of them. If, on the other hand, we do not presume to know where to draw the boundary between different food discourses – if we turn this boundary drawing into a central empirical question for the project – then network analysis becomes useful as a way to show emergent patterns in the way food is being talked about. What is at stake for the analysis is the ability to claim, based on very large volumes of text, that some chefs are talked about in qualitatively different ways than others, not just that they are talked about in certain places.

Digital methods and digital food studies

The field of digital methods has evolved over the past 20 years as new kinds of natively digital empirical material have become available and new computational techniques for exploring large volumes of unstructured data like text and images have become more accessible (Manovich, 2011; Rogers, 2019). One of the characteristic features of this development is the idea that data intensive analysis is no longer the remit of quantitative methods alone. Indeed, the combined ability to find patterns in messy data without discarding the rich information that allows researchers to make deeper forays into the context and meaning of a statement, a link, a "like," or other seemingly shallow digital traces has opened up new avenues for asking and answering qualitative questions with the aid of relevant computational techniques (Blok & Pedersen, 2014; Munk, 2019).

Besides a more general sociological/anthropological interest in bridging the quali-quantitative divide (see also Latour et al., 2012), digital methods have typically been advanced as a way to study how issues are debated online (Venturini, 2012; Marres & Moats, 2015; Burgess & Matamoros-Fernández, 2016). With a few exceptions (e.g., Munk & Ellern, 2015; Munk et al., 2016), food-related issues have mostly been absent from this literature, and yet it is a literature that is highly relevant to digital food studies. Indeed, there is perhaps no other topic so photographed and discussed online as food and the fact that this material is natively digital (i.e., brought into being on digital platforms) as opposed to digitised (e.g., scanned documents or images) produces a particular set of methodological challenges that online issue mappers have already been facing and dealing with for a couple of decades (for an early example, see Marres & Rogers, 2000). Any digital food study that engages with debate or discussion from online sources can therefore productively look to issue mapping and digital methods for inspiration.

One of the key challenges, that is shared between digital food studies and online issue mapping, is our reliance, as researchers, on the way platforms organise and make data available. We are implicitly or explicitly soliciting methodological decisions to the architects of these platforms all the time. The case I introduce below relies on two types of natively digital material, namely posts from pages on Facebook and names from lists and categories on Wikipedia. Just to take the former: A Facebook page is meant to be the public profile for a business or an organisation, as opposed to your personal profile or a group that you can be a member of. A page can have an address and a geolocation, it can be categorised by its owner according the kind of business it is doing, and it is allowed to "like" other pages. A page can post content, typically images, links, or videos accompanied by text, and it can allow its followers to comment or react to that content. If I want to find patterns in the way food actors on Facebook cite chefs, and especially if I want to do it at scale and therefore with some level of automation, I have to turn these features of the medium into affordances for the research I am doing. Rather than trying to come up with my own definition of what counts as a food actor and what does not (i.e., who should be included in the data set), I might as well make creative use of the page categories that are already in the data and thus ensure that my research design can actually be operationalised within the affordance space of Facebook. Similarly, rather than having an ambition to understand how these food actors really feel about their gastronomic heroes, it makes more sense to ask a question that recognises how Facebook pages are in fact used, for example, as strategic marketing tools where we will be able to see who food actors find it opportune to associate with.

Noortje Marres encourages us to acknowledge, and think critically with, this distributedness of methods (Marres, 2012). Digital media are not just archived online interaction, but methods in their own right. They nudge our informants to perform in certain ways, suggesting to them paths to be taken (and paths not to be taken), offering them ways to express themselves and interact with each other (while discouraging other ways to do so). We cannot ignore or work around this already existing methods ecology. Richard Rogers therefore mobilises the notion of "repurposing" to remind us that we should never uncritically rely on whatever the medium wants us to do, but rather creatively reuse digital media for purposes that fit our research interests (Rogers, 2018). There is of course no point in *re*purposing anything if the original purpose of that thing is not properly understood. It follows that doing research *with* data from the web invariably involves doing some research *on* and *about* parts of the web itself. As Rogers puts it, we have to be able to "follow the medium" (Rogers, 2009b). Marres makes a distinction between "discursivist" and "empiricist" approaches, the latter embodying the commitment to unpack how digital media platform make a difference to issues rather than simply following what is being said by different actors about these issues (Marres, 2015).

Building a corpus of Facebook posts from food-related pages

The first task of any digital methods project is to be build a data set or a "corpus." In this case, I needed a corpus of posts from food-related Facebook pages, and to some extent, I already had one. By casually browsing from one post in my feed to the next, I had manually collected a list of almost 200 pages that I thought could be interesting for the project to look at (in the sense that they frequently mentioned chefs when they posted, including the occasional Nordic chef). But since this collection was based on what was presented to me in my feed and hence heavily influenced by my actions as a Facebook user, I needed a better way to find pages. Besides, I wanted a larger corpus to have a better chance of detecting patterns in the way chefs were cited together. This poses a problem since Facebook, like most other digital platforms, does not offer a collection of carefully curated material for research. It is not an archive where you can know the principles of collection and search it accordingly. Neither is it possible to know the full population of anything, and so even if it was possible to systematically search for food related pages, proper sampling would still be impossible since we would not be able to know what it was a sample of. Again, this is why Richard Rogers calls social media "post-demographic machines" (Rogers, 2009a) and why repurposing is important. If sampling and representativeness, at least in any demographic sense of the word, is impossible, then what can we do within the possibilities offered to us by the platform?

As mentioned, owners of a Facebook page are allowed to "like" other pages on behalf of their own page. This feature can be repurposed for a "snowball" strategy where you start data collection with a small seed of material that is allowed to point you to more relevant material, in this case a seed of pages pointing to other potentially relevant pages through "likes." The seed can be very subjective and idiosyncratic – remember that my list of 200 pages was compiled relatively intuitively from browsing posts that I found interesting in my feed – but as the snowball picks up speed this idiosyncratic starting point is gradually counterbalanced as the question of what counts as interesting material is increasingly outsourced to the informants (in this case to more and more Facebook pages). It is possible to experiment with several snowballs from different starting points and compare when the corpora begin to converge.

In order to iteratively ask a platform like Facebook how an evolving collection of pages provide "likes" to other pages it is necessary to interact with the Application Programming Interface (API). This is the method through which third party applications pull data from the platform and hence another way in which we repurpose the platform. APIs are everchanging machines designed for other purposes than academic research, which makes it hard to replicate data sets, but it is normally the best option available to researchers who want to work with social media data (Lomborg & Bechmann, 2014). Most of the popular tools for data harvest are therefore also API-based. The corpus I have built for this project was constructed through the publicly available API endpoints in July 2018. To automate the process, I built a so-called script in the programming language Python. A script is a set of conditional instructions that tells the computer what to do in different situations. My script did the following:

1. Load the seed list of 200 pages.
2. For each of the seed pages, call the API and ask what other pages it likes.
3. For each of the newly found liked pages, decide if it is food-related and should be included in the corpus. For this task, I repurposed the page categories which allow page owners to self-classify under themes like "Chinese Restaurant," "Organic Grocery Store," or "Dairy Farm." I manually went through all the page categories that were available on Facebook at the time, identifying 317 of them as food-related. The script calls the API to ask for the page category, checks if it belongs to the 317 food-related categories, and, if it does, includes the page in the corpus.

4. Iterate the process on the next layer of the corpus: for each of the food-related pages liked by a page on the seed list, call the API, ask for page likes plus categories, and include in corpus if relevant.
5. Keep iterating and add new layers to the corpus.

I kept running the script until I had accumulated a corpus of 242K Facebook pages from around the world that I knew belonged to a food-related category. This was a practical cut-off point as the project was running out of time, but in principle, the snowball could have continued. I then called the API to ask for all the posts made by each of the pages in the corpus. The result is a collection of 102M text documents in different languages. In the table below, I provide a schematic overview of my protocol for all the steps in the project and suggest what you could do differently at each step (Table 15.1).

Building a query design for Nordic and international chefs

Having built the corpus, I now have to find a way to record when a chef is mentioned. Given that the size of the corpus does not permit manual coding, I have to do this detection automatically. This task would be the same regardless of how the corpus had been constructed (from Instagram, Twitter, lifestyle magazines, news media, cookbooks, whatever) as long as the time required to read and code the material exceeds what is possible inside the resource constraints of the project. It would also be the same challenge if I had been looking for ingredients, place names, or cooking techniques instead, although, as we shall see, the fact that personal names are usually spelled the same way in different languages means that more solutions are available for chefs than for, say, ingredients.

Essentially, there are two options. The first is to use a type of *Natural Language Processing* (NLP) called *Named Entity Recognition* (NER) to automatically detect personal names in the text. This option comes with the benefit of being able to discover bottom-up who is cited by the pages, but with the disadvantage of not knowing which of these discovered people are chefs. The list of extracted names would therefore have to be manually curated (again an impossible task considering the size of the corpus). Had I been looking for ingredients or place names instead, there are good models available. NLP also works to different standards in different languages (a consequence of differences in the volume and quality of available training data) and since I am working with a multilingual corpus, I should reasonably expect the solution to disadvantage food talk in smaller languages.

The second option is to compile a list of chef names that can be used as search terms to query the corpus. This comes with the advantage of knowing in advance that the names on the list are indeed chefs, and potentially, depending on the information available to the list builder, different kinds of metadata such as which country the chefs are from, what style of food they cook, etc. Since this option relies on a simple search, rather than language technology, it is also less sensitive to language

TABLE 15.1 Four-step research protocol

Step	*What I did*	*What you could also do*
1. Build a text corpus	*Collected 102M Facebook posts from 242K food-related pages found by snowballing page likes from a seed list and evaluating if page categories pertained to food.*	*Collect Facebook posts from pages you already know to be interesting (no need for a snowball). You could use a tool like the FacePager which is free and easy to use.* *Collect other kinds of online food talk such as tweets (you could use the TCAT tool), instagrams (you could use InstaLoader), or Reddit threads (you could use 4CAT).* *Collect other kinds of documents, such as magazine articles or scientific papers. Any kind of textual corpus will work with the next steps in the protocol.*
2. Query the corpus	*Collected names from lists of chefs on Wikipedia and searched the corpus for mentions of each of them.*	*Curate your own list of names that you know are interesting.* *Collect names from different lists of chefs or other food personalities.* *Use natural language processing to recognise names from the text in the corpus.*
3. Find patterns	*Built a network of chef names connected to each other if mentioned by the same page. This was done in Python using NetworkX. Then used a force vector layout algorithm in Gephi to find clusters of chefs that tend to be mentioned together.*	*Build a network of pages connected to each other if the mention the same chefs. This would make it possible to group the pages rather than the chefs.* *Build a network of pages connected to the chefs they mention. This would make it possible to explore directly in the graph visualisation which pages mention a particular chef or which chefs are mentioned by a particular page.* *You could use Table2Net as a user-friendly alternative to building networks in Python.*
4. Interpret the results	*Qualitatively analysed each of the clusters.*	*Although you can extract quantitative metrics for any given network, a qualitative analysis of clusters, centres, and bridges will probably be more interesting for your research interests.*

differences. Once the name of a chef is known, it does not have to be translated (although non-Latin alphabets still pose a challenge). The drawback, however, is that we have to decide *a priori* who are relevant chefs to search for rather than soliciting this task to the food actors we are studying on Facebook.

Given the practical challenges associated with the NLP approach, I am, in this case, going with the simpler query option (but with a smaller English language corpus, the benefits of the NLP approach could likely have outweighed its disadvantages). The next step is therefore to acquire a list of chefs to work from, and this is where it becomes necessary to repurpose Wikipedia. As it turns out, getting a good list of chefs is more difficult than you might think. There are national and international chef associations, but they typically do not list their members online, and, if they do it, it is debatable whether these members are in fact the top tier chefs that would be cited in online food talk. There are also privately curated lists of chefs. San Pellegrino, for instance, publishes a ranking of the best chefs in the world, but I suspect that it is itself part and parcel of a very distinct culinary discourse already. Wikipedia, on the other hand, offers what appears to be a quite encompassing list of chefs from different centuries (https://en.wikipedia.org/wiki/List_of_chefs). I am here scraping it for names under the twentieth and twenty-first century headings.

Believing that Wikipedia is not a perspective, however, would be a mistake. The point about repurposing is still valid here. On the "talk" page behind the list (https://en.wikipedia.org/w/index.php?title=Talk:List_of_chefs) the editors are discussing the criteria for inclusion. Who should be counted as a chef? Who is important enough to be on the list? It turns out that the original version of the list was copy pasted from French Wikipedia and since elaborated. It is this online community of Wikipedians and their ideas about good list building that I am now repurposing as part of my method. The same is true for the categories of Norwegian (https://no.wikipedia.org/wiki/Kategori:Norske_kokker), Swedish (https://sv.wikipedia.org/wiki/Kategori:Svenska_kockar), and Danish chefs (https://da.wikipedia.org/wiki/Kategori:Kokke_fra_Danmark) from which I compile the Nordic part of the query design.

Building a co-chef network

The process of turning the lists of names from Wikipedia into a network of chefs cited by the same pages is essentially a very large search operation. For each of the 683 chefs, I am searching the text of each of the 102M posts in the corpus to see if the chef is mentioned in the actual post text on Facebook. I am using another Python script for this purpose. Whenever two chefs are mentioned in posts by the same page, the script registers a link between them. The script also associates each link with a weight that becomes heavier as more pages make the same connection. In network parlance, I am representing the chefs as "nodes" and the links as "edges" (see Figure 15.1), which simply means that I have converted the results from the query to a relational data structure where chefs are always related with different weights to one another. Using the Python library NetworkX, I am able to export the network as a .gexf file that can be opened in Gephi and explored visually. Gephi is an open-source tool for explorative network visualisation (Bastian et al., 2009) and can be downloaded freely at https://gephi.org/. It allows easy layout and manipulation of graphs, such as sizing, colouring, and filtering nodes and edges, as well as placing them visually in space. As a user, you are encouraged to experiment with these settings and explore their implications.

In order to keep visual track of my Nordic chefs, I am giving my nodes a triangular shape (useful when colour is not an option). I then apply a layout algorithm called ForceAtlas2 (Jacomy et al., 2014) that helps me spot clusters of chefs that are frequently cited together. A "force vector" or "spring-based" layout works as a repulsive force that pushes all nodes apart from each other unless they are held together by an edge (i.e., edges act as springs). In Figure 15.1, I am spatialising a smaller sample of the network with ForceAtlas2 to demonstrate how nodes become visually grouped into clusters. Figure 15.2 also shows how the network is constructed: in a situation where FB Page 1 mentions Chef X and Chef Z, these two chefs will be represented as nodes connected by an edge. When FB Page 2 then also mentions both Chef X and Chef Z, the weight of the edge between the nodes increases.

The network sample in Figure 15.1 can be seen as a proof of concept that force vector spatialisation, that is, the process of visually pushing nodes apart in the layout unless they are connected by edges, allows us to see relevant clusters of co-cited chefs. Two such clusters can be clearly delineated. Most of the Nordic chefs (triangular nodes) tend to be cited together by some pages, while all the international chefs (circular nodes) tend to be cited together by other pages. All the chefs in the

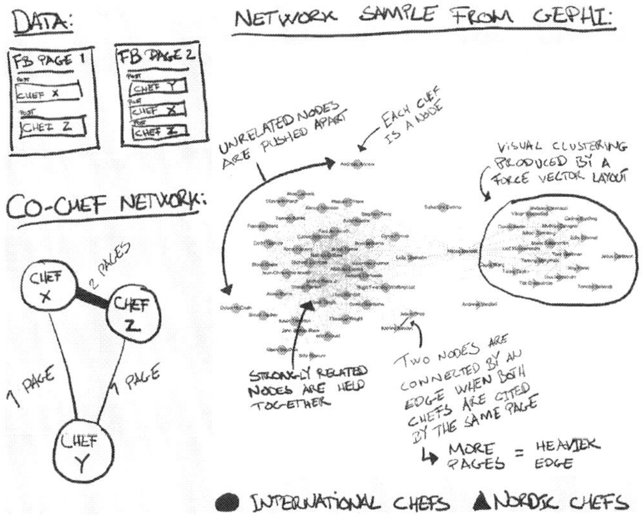

FIGURE 15.1 First concept of how to build the co-chef network and a sample testing the idea in Gephi.

Nordic cluster to the right are Swedish which would suggest that they are part of a national culinary discourse. This is further reinforced by the fact that most of the Nordic chefs found outside this cluster (Katrine Klinken, Silla Bjerrum, and Adam Price) are Danish. Leila Lindholm, who is Swedish, clusters with the international chefs, which would suggest that she has become part of a different culinary discourse than her countrymen in the sample, a reasonable interpretation given that she has had her cookbooks translated into multiple languages and has also appeared in cooking shows on the BBC.

Although the force vector layout places most of the Swedish chefs in close proximity to one another, their nationality never has an influence on the layout. The only information the algorithm relies on to produce the layout is the relations between the nodes. When most Swedish chefs in the sample end up in the same cluster it is simply an expression of the fact that they tend to be cited by the same pages. Rather than assuming that nationality determines the way chefs belong to culinary discourses, which would prevent us from seeing how a chef like Leila Lindholm belongs differently, the network approach allows us to map discourses relationally. In fact, since nationality is not a factor in the clustering, we can also use the visualisation to identify chefs that have been miscategorised by Wikipedia. Malin Söderström, for instance, is Swedish, but the Wikipedians who curated the lists from which I got the names in the first place did not recognise her as such. Yet, based on the way she is cited alongside other Swedish chefs, the force vector layout is able to place her in the cluster emerging as a distinctly Swedish culinary discourse to the right in Figure 15.1.

Analysing the co-chef network

The spatialised sample thus promises a method for delineating culinary discourse relationally and from the bottom up. However, as we move from the sample to the full network it becomes clear that additional manipulation is necessary for the force vector layout to keep showing clusters. In Figure 15.2, which is an unfiltered version of the full network, we still see some smaller clusters on the right but the vast majority of the nodes are grouped in a densely connected "hairball" where it is possible to discern some regions (nodes on the left side of the hairball seem more related to each other than to nodes at the bottom or at the top of the hairball) but not at all clear where to delineate distinct discourses. In network terms, this hairball can be explained by the fact that many of the nodes are connected to almost all other nodes in the graph. When we run a force vector layout, the edges of these highly connected nodes act as springs that prevent the rest of the nodes from being pushed apart. In Figure 15.2, I have sized the nodes by their "degree" (a measure of how many connections they have) to make the effect clear.

Translated to our data set, we can surmise that some chefs are cited so widely that they have also been cited, at some point, by the same page as almost any other chef on our list. There is no particular pattern to the way these omnipresent chefs

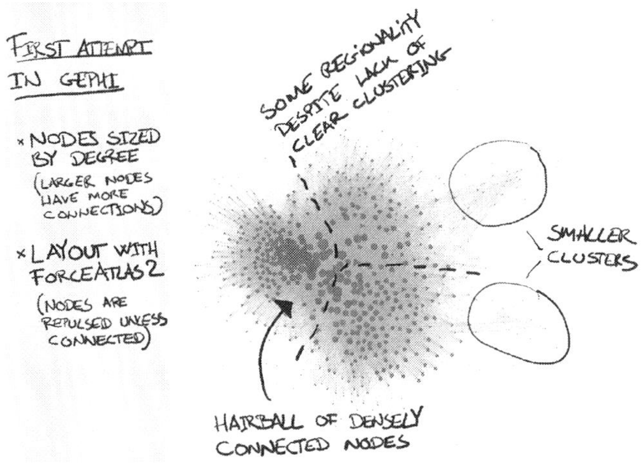

FIGURE 15.2 The full network of 669 chefs connected by co-occurrence on 242K food pages analysed in Gephi.

are being talked about. Or more precisely, there is no particular pattern to the company of chefs they are being talked about in. It will become clear later that such patterns among chefs do indeed exist if we disregard the most omnipresent among them – there are very clear spheres of food talk across our 242K Facebook pages, each of them characterised by a distinct universe of cited chefs – and so the fact that some chefs are so widely talked about that they transcend this pattern constitutes a finding in itself.

In Figure 15.3, I have produced a so-called "ego network" for the three chefs that co-appear with most other chefs in the network. An ego-network is a subset of the graph consisting only of the nodes connected directly to the node in question (in our case, the chefs that are mentioned on the same pages as the chef in question). The first thing to notice here is that there is almost no noticeable difference between these subgraphs and the full graph shown in Figure 15.2. It makes very little difference to the graph if we introduce a filter that excludes all chefs that have not been cited on the same page as Jamie Oliver, Gordon Ramsay, or Paul Bocuse.

This is a good time to remember why we are doing visual network analysis in the first place. Typically, and especially when we are working with force vector layouts, the idea is to discover a topology where clusters of connected nodes can be distinguished from one another. In our case, the idea is to spot clusters of chefs that tend to be cited by the same pages. Such clustering tendencies in complex relational data can be hard to make visible, but force vector layouts are designed to help us do it. They are not, however, particularly helpful if we want to spot the top 100 most connected nodes. These nodes are much better shown on a simple list where the items can be ranked. In fact, trying to retain this information as part of a network visualisation can actively prevent us from appreciating the underlying topology (as

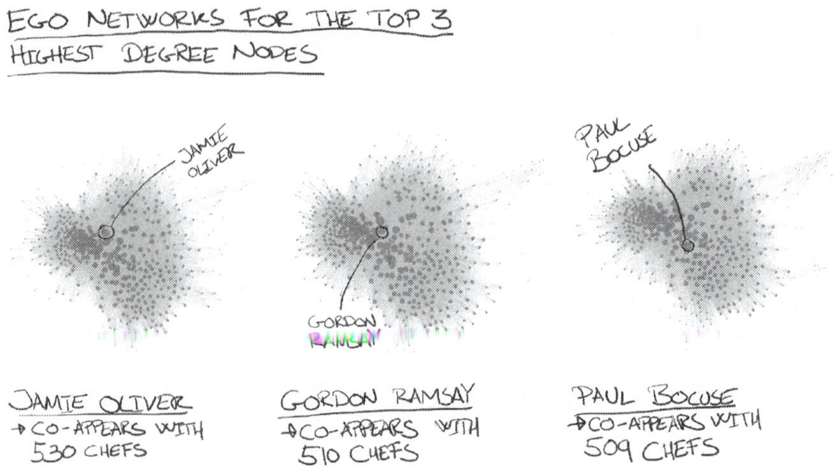

FIGURE 15.3 Exploring the neighbourhoods of the most omnipresent chefs and the effects of these nodes on network topology.

we have just seen above). Instead, we could just produce a list of the high degree chefs, like so:

1. Jamie Oliver (co-cited on pages with 530 other chefs)
2. Gordon Ramsay (co-cited on pages with 510 other chefs)
3. Paul Bocuse (co-cited on pages with 509 other chefs)
4. James Beard (co-cited on pages with 490 other chefs)
5. Thomas Keller (co-cited on pages with 486 other chefs)
6. Alain Ducasse (co-cited on pages with 481 other chefs)
7. Julia Child (co-cited on pages with 475 other chefs)
8. Ferran Adriá (co-cited on pages with 471 other chefs)
9. Massimo Bottura (co-cited on pages with 457 other chefs)
10. Daniel Boulud (co-cited on pages with 456 other chefs)
11. Heston Blumenthal (co-cited on pages with 451 other chefs)
12. René Redzepi (co-cited on pages with 446 other chefs)
13. Anthony Bourdain (co-cited on pages with 436 other chefs)
14. Eric Ripert (co-cited on pages with 425 other chefs)
15. …

Now that we know that these chefs are very widely talked about across almost any context, we can freely filter them out of the visualisation. There is, in fact, nothing more an analysis of the network topology can teach us about them. In Figure 15.4, I therefore experiment with different thresholds for such a filter, subsequently redoing the force vector layout for the remaining graph in each instance (this is easily done in Gephi). A topology with distinct clusters gradually emerges as more high

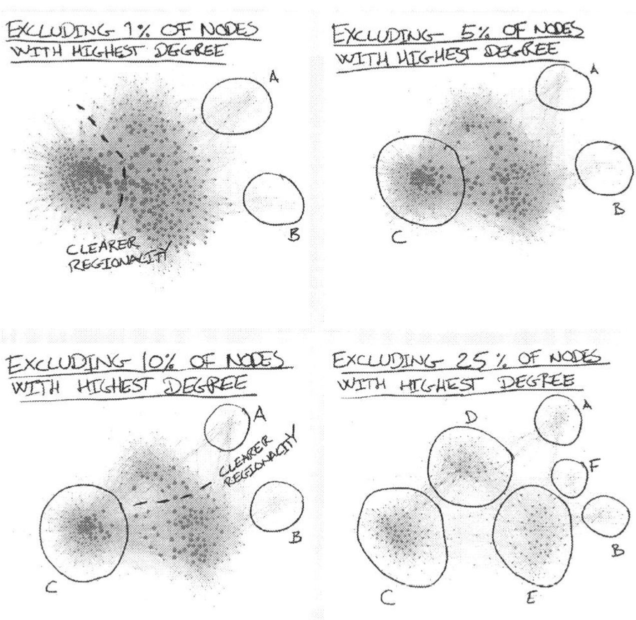

FIGURE 15.4 Experimenting with different thresholds for a filter based on node degree.

degree nodes are filtered out. With 75% of the original graph visible, six clusters can be clearly delineated.

Eventually, through trial and error, I settle for a filter that removes the 20% highest degree nodes from the graph (see Figures 15.5 and 15.6). This setting gives the clearest clustering (again, the point of removing the highest degree nodes is to see clustering) with the least loss of information (the point of not filtering too heavily). It remains, however, a qualitative cut that could justifiably have been made differently. As I lower the threshold for the degree filter, the chefs that are left in the graph become less and less omnipresent. The graph produced by removing the top 10% highest degree nodes, for example, has three clear clusters and a large hairballed centre from which the remaining three clusters will emerge when we lower the threshold further. Some of the chefs that produce this hairball and get filtered out with a lower threshold (Atul Kochhar, Tom Kitchin, or Angela Hartnett, for instance) only have strong connections across two or three of the eventual six clusters. They are not omnipresent in the same way as Paul Bocuse, Jamie Oliver, or Gordon Ramsay (the highest degree chefs). Choosing a different filtering threshold would have allowed me to explore these differences in omnipresence further. Indeed, using an entirely different filter, such as the betweenness metric which measures the degree to which a node is on the shortest path between all other nodes in the network, I could have reoriented the analysis towards chefs that bridge two or three otherwise unrelated clusters without necessarily being omnipresent in all food talk on Facebook. No matter how I set the filter, though, I want to get to a point where I can meaningfully interpret the resulting topology. Having set the threshold at 80% I am left with six

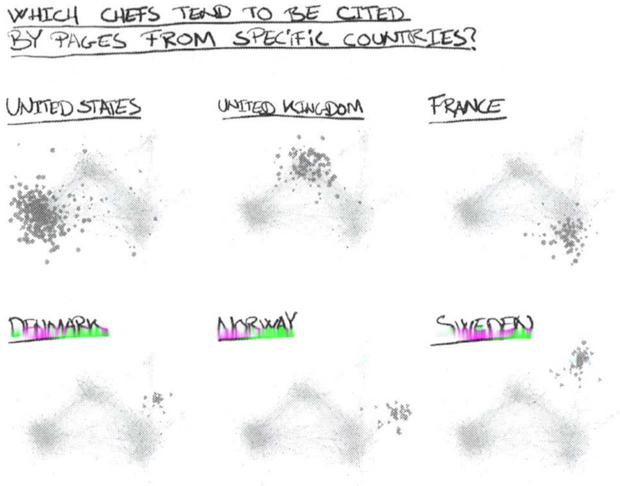

FIGURE 15.5 Interpreting the network topology: Which chefs are mentioned a lot by pages from different countries.

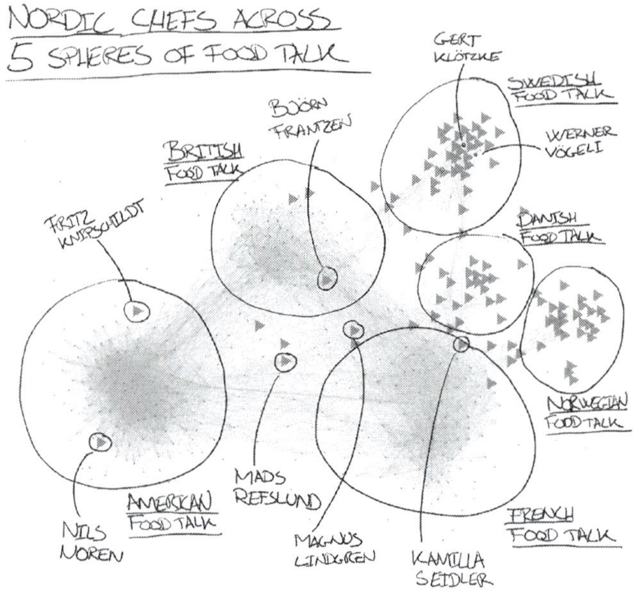

FIGURE 15.6 Interpreting the network topology: Annotated clusters and nodes, Nordic chefs enlarged as triangular nodes.

clearly delineated clusters. The question is why? What could be the reason why the chefs on my list tend to be mentioned together in this particular pattern across the 242K food pages in my data set?

At this point, I can either dive straight into the chefs found in the different clusters, search for the textual contexts in which they are mentioned in my data set, read about their background online, and thus try to stitch together a narrative that makes sense. Qualitative work, in other words. Or, if it is available to me, I can try to project some additional information onto the network, for instance by sizing or colouring the nodes according to some variable. In this case, I know the country where each of the 242K food-related pages comprising the data set are located geographically (my script got this information from the Facebook API when it was snowballing). I was therefore also able, when I was searching through my list of chefs, to count how much each chef is mentioned by pages from each country. In Figure 15.5, I have projected some of this information onto the network by sizing the nodes according to how much they are cited by pages from six different countries.

The result provides a good clue to an interpretation. Two of the larger clusters are dominated by chefs that tend to be cited by American and British pages, respectively. This does not prevent the chefs in these clusters from being cited by pages from other countries as well, but the fact that the clustering corresponds to a geographical delimitation tells us that American and British pages produce two different patterns of co-cited chefs, that is, two distinct culinary discourses. The same is true for Danish, Swedish, and Norwegian pages, which produce the three smaller clusters on the top right. If we take a qualitative look at the chefs comprised in these five culinary discourses it turns out that the vast majority of them are also American, British, Swedish, Norwegian, and Danish nationals, respectively. In other words, when we filter out the chefs that everyone talks about in any context, it seems that Danish pages talk about Danish chefs only, Swedish pages about Swedish chefs only, and so forth. These five culinary discourses are essentially reflective of five national food scenes.

The sixth cluster at the bottom right, however, is more difficult to categorise. Parts of it correspond well to food talk on French pages, but not all of it. Indeed, pages from countries like Germany, Italy, Spain, or Switzerland are equally contributing to this cluster (although not shown in Figure 15.5). At closer inspection, the chefs in this cluster also have different and not just European nationalities. There is, in other words, no national food scene here. What we are dealing with seems to be a sort of international gastrosphere where the entry ticket is a Michelin star or similar.

With an interpretation of the topology in place (Figure 15.6), I can now explore the question about Nordic chefs that I was interested in to begin with. I already know from my degree filter (Figure 15.4) that some of the Nordic chefs have achieved so wide international recognition that they are mentioned across contexts and in the company of so many other chefs that they transcend any particular culinary discourse. These omnipresent Nordic chefs are Claus Meyer, René Redzepi, and Rasmus Koefoed (Denmark), Magnus Nilsson, Marcus Samuelsson, and Emma Bengtsson (Sweden), and Esben Holmbo Bang (Norway, although a Danish national).

Conversely, I can also assert that a good portion of the remaining Nordic chefs on my search list are not recognised outside the national culinary discourses in their respective countries. Some of them, like Frantz Knipschildt, who as a young chocolatier began his own production in New York, became widely renowned in the United States and never returned to a career in Denmark, are virtually unknown in their country of origin. Magnus Lindgren also made his career with Heston Blumenthal in the UK and was never on the gastronomic radar in Sweden. Others, like Niels Norén, Björn Frantzén, or Mads Refslund, have moved abroad to spearhead the New Nordic Cuisine movement with restaurants in places like New York or Hong Kong after claiming fame in Stockholm and Copenhagen first (Refslund was the co-founder of Noma). Something similar should be the case for Kamilla Seidler, who was named the best female chef in Latin America after she opened her New Nordic restaurant in La Paz, but that seems to be a story that has mostly gained attention on the fringes between the national Danish discourse and the French/Michelin discourse.

Conclusion

The mapping thus allows me to show how chefs are talked about in separate spheres online; spheres that *turn out to be* predominantly but not entirely national in character. Rather than claiming *tout court* that there is a Danish, Swedish, and Norwegian culinary discourse in which online tribute is paid to Danish, Swedish, and Norwegian chefs, Figure 15.6 both urges and supports a considerably more flexible analysis. Although an overall pattern of separate national spheres is clearly visible, we are forced to reckon with each individual chef and their different ways of belonging in this relational space. The nationality of a Nordic chef does not determine this belonging, that much is clear from Figure 15.6, and the same is of course true for non-Nordic chefs. Werner Vögeli, who is Swiss but has had a career in Sweden, is thus clearly part of the Swedish national discourse.

The strength of visual network analysis is that it does not presume any prior categories, except for the kind of relationality we choose to include as edges between the nodes. We can read the map in Figure 15.6 as a way to categorise the nodes based on the way they relate to one another rather than the predicates we have given them in advance. In our case, this means that we can read the map as a way to categorise chefs based on how they are part of the same discourses (mentioned by the same food-related pages) rather than their nationality, their style of cooking, their age, or any other *a priori* category. In fact, once we have discovered the emergent categories represented by the clusters in Figure 15.6, and once we have concluded on the basis of our research design that these categories are in fact different discourses, we can use the network to discuss the role of nationality in this discursive landscape. We can use it to become curious about individual chefs who break a pattern or about clusters that are hard to explain. This is a fundamentally explorative process that is useful for setting a direction for further qualitative work. It is not a way to confirm or refute a hypothesis with some degree of probability. We are using quantitative techniques to do qualitative ground work on volumes of messy data that would otherwise be impossible to handle.

Now you try

Download the network of co-cited chefs from https://github.com/akmunk/ediblenorth/blob/master/ChefNetwork.gexf and install the latest version of Gephi from (www.gephi.org). Good tutorials are available on the website. Open the .gexf file and run the ForceVector2 layout algorithm (adjust the scaling if you need to separate the nodes further). From the filters tab, select a degree filter and try to exclude the highest degree nodes. Run the layout algorithm again and notice how clustering becomes clearer.

Questions for reflection

How could your research problem be explored by visually clustering some aspect of your data set with a force vector layout?

What could be the nodes? And what kind of relationships could define the edges between them?

How could you justify filtering the graph to make clustering clearer?

How could you repurpose digital media to build a data set or make interesting query designs?

Ethical considerations

Internet research is riddled with ethical concerns, not least in relation to personal and sensitive information, and as a consequence of the fast pace of the political debate on data ethics the goal posts of are constantly moving. This is felt when platforms like Facebook decide to shut down large parts of their data access in response to GDPR, or when machine learning models become capable of predicting sensitive information, such as political or sexual orientation, from a few, innocent likes on Instagram or the writing style of a tweet. Some guidelines are available through the Association of Internet Researchers (https://aoir.org/reports/ethics2.pdf).

TABLE 15.2 Technical vocabulary

API (Application Programming Interface): A technology that makes it possible to for a computer to ask a website for certain types of information. Different social media platforms have different APIs with their own rules and syntax.

Force vector layout: A way to find clusters in network by visually placing the connected nodes close to each other. Works by introducing a repulsive force that push all nodes apart from each other, unless the nodes are connected by edges which act as springs holding the nodes together.

Gephi: A piece of open-source software for doing visual network analysis.

Network: A relational data structure consisting of nodes and edges. Nodes can be any type of entity and edges any type of relation between these entities. For instance, authors (nodes) related by the way they cite (edges) each other in scientific journals; or, hashtags (nodes) related by the way they co-occur (edges) in tweets.

Script: A piece of code, typically written in a programming language like Python, R, or JavaScript, that makes it possible to automate instructions to a computer and make these instructions conditional on various criteria being satisfied.

Further readings

Rogers, R. (2019). *Doing Digital Methods*. London: SAGE Publications Limited.

The tool repository of the Digital Methods Initiative in Amsterdam: https://wiki.digitalmethods.net/Dmi/ToolDatabase.

The blog of Martin Grandjean (http://www.martingrandjean.ch/) for good tutorials.

The blog of Tommaso Venturini (http://www.tommasoventurini.it/wp/) for use cases and reflections.

The Gephi community on Facebook (https://www.facebook.com/groups/gephi/) for peer group and support.

References

Bastian, M., Heymann, S., & Jacomy, M. (2009). *Gephi: an open source software for exploring and manipulating networks*. In *Third International AAAI Conference on Weblogs and Social Media*, San José, CA.

Blok, A., & Pedersen, M. A. (2014). Complementary social science? Quali-quantitative experiments in a Big Data world. *Big Data & Society*, 1(2), https://doi.org/10.1177/2053951714543908.

Burgess, J., & Matamoros-Fernández, A. (2016). Mapping sociocultural controversies across digital media platforms: One week of # gamergate on Twitter, YouTube, and Tumblr. *Communication Research and Practice*, 2(1), 79–96.

Høyrup, J. F., & Munk, A. K. (2007). Translating terroir: Sociomaterial potentials in ethnography and wine-growing. *Ethnologia scandinavica*, 37, 5–20.

Jacomy, M., Venturini, T., Heymann, S., & Bastian, M. (2014). ForceAtlas2, a continuous graph layout algorithm for handy network visualization designed for the Gephi software. *PloS one*, 9(6), e98679.

Latour, B., Jensen, P., Venturini, T., Grauwin, S., & Boullier, D. (2012). 'The whole is always smaller than its parts' – A digital test of Gabriel Tardes' monads. *The British journal of sociology*, 63(4), 590–615.

Lomborg, S., & Bechmann, A. (2014). Using APIs for data collection on social media. *The Information Society*, 30(4), 256–265.

Manovich, L. (2011). Trending: the promises and the challenges of big social data. *Debates in the digital humanities*, 2, 460–475.

Marcus, G. E. (1995). Ethnography in/of the world system: the emergence of multi-sited ethnography. *Annual Review of Anthropology*, 24(1), 95–117.

Marres, N. (2012). The redistribution of methods: on intervention in digital social research, broadly conceived. *The Sociological Review*, 60, 139–165.

Marres, N. (2015). Why map issues? On controversy analysis as a digital method. *Science, Technology, & Human Values*, 40(5), 655–686.

Marres, N., & Moats, D. (2015). Mapping controversies with social media: the case for symmetry. *Social Media+Society*, 1(2). https://journals.sagepub.com/doi/full/10.1177/2056305115604176#articleCitationDownloadContainer

Marres, N., & Rogers, R. (2000). Depluralising the Web, repluralising public debate-the case of the GM food debate on the Web. In Rogers, R. (ed.), *Preferred Placement: Knowledge Politics on the Web* (pp. 113–136). Maastricht: Jan van Eyck Akademie Editions.

Munk, A. K. (2019). Four styles of quali-quantitative analysis: making sense of the new Nordic food movement on the web. *Nordicom Review*, 40(s1), 159–176.

Munk, A. K., Abildgaard, M. S., Birkbak, A., & Petersen, M. K. (2016). *(Re-) Appropriating Instagram for social research: three methods for studying obesogenic environments.* In *Proceedings of the 7th 2016 International Conference on Social Media & Society* , London, pp. 1–10.

Munk, A. K., & Ellern, A. B. (2015). Mapping the new Nordic Issuescape: how to navigate a diffuse controversy with digital methods. In Jóhannesson, G.T., Ren, C., & van der Duim, R. (eds.), *Tourism Encounters and Controversies: Ontological Politics of Tourism Development* (pp. 73–96). London: Routledge.

Munk, A. K., & Jensen, T. E. (2015). Revisiting the histories of mapping. *Ethnologia Europaea*, 44(2), 31.

Rogers, R. (2009a). Post-democraphic machines. In Dekker, A. & Wolfsberger, A. (eds.), *Walled Garden* (pp. 29–39). Rotterdam: Virtueel Platform.

Rogers, R. (2009b). *The End of the Virtual: Digital Methods* (Vol. 339). Amsterdam: Amsterdam University Press.

Rogers, R. (2018). Otherwise engaged: social media from vanity metrics to critical analytics. *International Journal of Communication*, 12(732942), 450–472.

Rogers, R. (2019). *Doing Digital Methods*. London: Sage.

Venturini, T. (2012). Building on faults: how to represent controversies with digital methods. *Public understanding of science*, 21(7), 796–812.

AFTERWORD

Future methods for digital food studies

Deborah Lupton

Food cultures have always been sites of diversity, political contestation, and intense affective forces, entangled with notions of embodiment, identity, family, and social relations (Caplan, 1997; de Solier, 2013; Lupton, 1996). The mass media have played an important role in generating and reproducing values, meanings and practices related to food production and consumption (Ashley, Hollows, Jones, & Taylor, 2004; Goodman, Johnston, & Cairns, 2017; Taylor & Keating, 2018). In recent decades, digital media have dramatically expanded these opportunities (de Solier, 2018; Lewis, 2018; Lupton, 2018a; Lupton & Feldman, 2020; Rousseau, 2012). Where once media coverage of food relied on a limited number of professionals: big-name chefs' cookbooks and television cooking programs, restaurant reviewers, food writers, and the like, new media formats have emerged since the turn of the twenty-first century which have revolutionised opportunities for amateurs to make and share their own content. These media include blogs, videos, social media sites, hashtags and other tagging practices, online comments, and curating platforms, all of which can potentially reach massive audiences worldwide.

As more portrayals of food production, preparation, and consumption go online across the world, a challenge is presented to social researchers as to how best to capture and analyse these practices and ideas. Each day, a huge volume of data about people's lives, preferences, and habits are uploaded to the internet. A 2019 infographic presented figures such as 277,777 stories and 55,140 images posted on Instagram, 4.5 million videos watched on YouTube, over 4.4 million Google searches, and 8,683 food orders received by GrubHub *every minute of the day* (Domo, 2019). Many platforms and apps these days are linked to each other using hyperlinks or other affordances for sharing content across sites. Algorithms and machine learning processes operate to shape users' experiences in ways that can be difficult to identify. There is a form of recursive mediation occurring in digital food media texts: food itself is medium for cultural meaning and communication, digital food texts remediate the medium of food, which in turn can influence practices of mediation, such as people's creation of content for social media or websites or their food purchasing and cooking practices, which they may choose to share on digital media sites … and the cycle flows on.

How can social researchers deal with these lively, responsive, and often very "big" food data? Perhaps most compellingly – at least from my perspective, how can the multisensory, political, moralistic, spiritual, and affective nature of food – its vitalities and visceralities (Lupton, 2019) – be accessed using social inquiry methods? What approaches can access and make sense of the diversity, dynamism, entanglements, and sheer volume of these portrayals as they constantly shift and change, responding to users and to each other?

In the recent collection I edited with Zeena Feldman, *Digital Food Cultures* (2020), most authors (including myself) adopted a critical discourse analysis or social semiotic approach to investigating digital foodscapes. Both approaches attempt to site media texts or responses to these texts in their broader sociocultural and political contexts. Their material included websites, YouTube videos, Instagram posts, Facebook pages, blogs, online news articles, promotional material by tech developers, discussion forums and comments, interviews with and diaries completed by research participants, and an autoethnographic "self-experiment." The contributors to our book selected a small amount of material from their chosen medium, using parameters such as choosing a limited time frame, the most popular outlets or focusing only on one or two "influencers," to allow for detailed analysis. These choices were led by their research questions. In some cases, authors were interested in analysing online content, while in others, they focused their attention on how people were engaging with content or with software such as apps.

The contributors to this book, which builds on and supplements *Digital Food Cultures* by focusing explicitly on research methods, have presented some further interesting ideas for ways forward. As some authors note in their chapters, traditional approaches still have much to offer the researcher who is interested in digital food cultures. In terms of the use of social and cultural theory, the dominant approaches in this volume, as in the *Digital Food Cultures* book, are textual analyses. In her chapter, Michelle Phillipov provides a useful overview of the different varieties of analysis that fall under this rubric. Netographic analyses of websites have been undertaken since the early years of the internet. The chapter by Meghan Lynch and Kerry Chamberlain adopting a netographic approach to health- and weight-loss-related food blogs, involving detailed observation and thematic analysis of the blogs' content, is another example of a well-used approach. To understand practices of digital self-tracking and representation in relation to food, Rachael Kent argues for a multimodal digital ethnographic approach that combines in-depth interviews with guided diary entries and screenshots provided from social media and app content by her study participants.

Several authors in this volume adopt a social justice or activism perspective in outlining the possibilities for digital food research methods. In her chapter, Tania Lewis addresses the question of how social research can be conducted into food activism as it takes place using digital media, with a focus on practices related to ethical consumption and sustainability. Jonatan Leer's study takes up an action research perspective based on the critical theory of Adorno and others, working collaboratively towards giving a voice to his research participants in the interests of activism and social change. Alana Mann draws attention to the role of hashtags, handles, and retweets in organising Twitter-based digital media activism about food issues, highlighting their rhetorical, intertextual, and interdiscursive dimensions.

Another set of chapters grapple with the challenge of massive, constantly changing, and ephemeral data sets. In her chapter, Caroline Nyvang shows how internet archives of food-related material can be a rich source of social inquiry. Challenging the oft-held notion that "the internet never forgets," Nyvang highlights the importance of understanding archiving logics and infrastructures to better preserve this material. Thomas Mosebo Simonsen and Stinne Gunder Strøm Krogager consider how to study food on YouTube. Adopting a computational method to work with large datasets, Anders Kristian Munk sought to identify the co-cited networks of well-known chefs on Facebook, using Gephi, an open-source graph visualisation tool. Camilla Vásquez and Alice Chik's chapter also provides an explanation of how to whittle down very large datasets (in their case, negative restaurant reviews in a specific geographic area published on Yelp and TripAdvisor) to enable a detailed linguistic analysis of the language used, providing suggestions for software that can help to automate this task.

For Leer, the digital element of his research is using a digital camera to record everyday meal practices in Danish day care centres and disseminating the findings on digital media. In his case, it is the researcher who is digitising the food practices under investigation and then using these images as research materials. A similar approach is adopted by Fabio Parasecoli, who describes his use of digital photography as a pedagogical tool for his food studies students. In addition to examining content on websites, review sites and social media accounts related to local (Brooklyn, New York City) hipster food practices and establishments, Parasecoli's students were asked to create digitised mood boards and post-relevant images to Instagram using the #globalbrooklyn hashtag, thus participating in creating digital media content. Lewis' discussion of a project on food activism also recounts the use of digital photography as a way of documenting practices as part of a digital ethnography.

A number of contributors draw attention to the volatility and dynamic nature of the interplay between human actors, digital platforms and the content they present. Katrine Meldgaard Kjær examines how social media "influencers" gain credence and shape fans' practices with an emphasis on the infrastructural dimensions of software, such as the affordances of website platforms. Other contributors looked at how people engaging with digital food media and technologies consume this content and the implications for their everyday practices, including but not limited to food consumption. Kent's study, for example, demonstrates how people use food-tracking media in practices of selfhood and embodiment.

Where to from here? While such phenomena as food activism using digital media (Schneider, Eli, Dolan, & Ulijaszek, 2018) and discussions and practices related to clean eating, the paleo diet, fitness and health, food waste, the promotion of local food cultures and most notably vegetarianism and veganism on websites, discussion forums, YouTube, and social media have begun to receive detailed attention (Lupton & Feldman, 2020), there is a range of popular digital media that has not yet been examined to any great extent by social researchers, including the visual cultures of emojis, GIFs, and memes. As Schneider and Eli point out in their chapter, novel food production technologies such as those offered by "smart farming" initiatives, involving digital sensors embedded in tractors, livestock and soil, are burgeoning, yet few social researchers have sought to investigate how they are used and what their unintended consequences may be. New social media sites such as Tiktok (particularly popular among teenagers at the time of writing) deserve more attention.

The interplay of digital technologies designed to enhance dining and other eating experiences has received extensive attention in experimental psychology and human–food interaction studies, but very little by researchers in media, communication, cultural studies, or sociology. Nicolai Jørgensgaard Graakjær notes the dominance of the visual and verbal over other sensory dimensions. His chapter focuses on the semiotics of the auditory dimensions of digital food media, such as ASMR videos and beverage advertisements. As Jørgensgaard Graakjær points out, the sounds food and drink make when they are processed, prepared, and consumed can be important to how attractive and tasty it is perceived as being, yet this is an almost completely neglected feature in digital food social research.

For me, thinking with theory offers an exciting approach that acknowledges that social inquiry (or indeed, any kind of research) is a research assemblage, a more-than-human coming together of humans with nonhumans. Consonant with this approach is recognition of the liveliness and responsiveness of digital representations (Lupton, 2016, 2017). Any form of culture and practice is subject to continual change, and must be understood in its historical, social, spatial, and political contexts. The acceleration and volume of digitised food practices and portrayals in some ways point to these vitalities and emphasise the inevitable emergent and partial nature of inquiry into cultures.

The intensities of engagement with digital food media, including app and other software use as well as content creation and responses to others' content, has not been acknowledged in past research to any great extent. Yet it is these affective flows – often involving powerful ambivalences and contradictions, and spanning feelings of anger, hatred, misogyny, racism, disgust to pleasure, and self-indulgence – that underpin the huge popularity of digital food media. In recent research, I have taken up more-than-human theory to attempt to understand the vitalities – or what Bennett (2009) refers to as "thing-power" – of digital food media. In a project investigating women's use of food tracking apps (Lupton, 2018b), for example, I drew on interviews to present vignettes of five women's experiences and the affective forces and agential capacities that their experiments with these apps had opened or closed. In my chapter in *the Digital Food Cultures* volume (Lupton, 2020), I examined affective dimensions of what I describe as "carnivalesque" food videos on two YouTube channels, both of which present strongly ambivalent and highly gendered portrayals of excessive eating practices.

Attempting to analyse digital food cultures as more-than-human requires new methods as well as different viewpoints and entry points. Lewis takes up this perspective in her chapter, arguing for incorporating research methods that recognise the other-than-digital elements of food media engagements and enactments, including technologies such as chainsaws and hammers as well as smartphones, and living creatures other than humans. More-than-representational approaches to social inquiry (Thrift, 2008; Vannini, 2015) can begin to access the thing-power of media in ways that go beyond what can be readily articulated in words.

In my Vitalities Lab, we have been experimenting with design- and arts-based methods which offer alternative approaches to social inquiry. These include zine making, drawing and mapping, and using creative writing prompts to stimulate people's thinking around research topics. These approaches have the potential to engender imaginaries, affects, memories, and speculations about food and its futures. When applied to digital food cultures, they can operate to emphasise the materiality of digital media and the entanglements between humans and nonhumans (organic and technological) as images, words, sounds, physical experiences, and sensations are mediated and remediated, shared, and spread.

References

Ashley, B., Hollows, J., Jones, S., & Taylor, B. (2004). *Food and Cultural Studies*. London: Routledge.

Bennett, J. (2009). *Vibrant Matter: A Political Ecology of Things*. Durham: Duke University Press.

Caplan, P. (Ed.). (1997). *Food, Health and Identity*. London: Routledge.

de Solier, I. (2013). *Food and the Self: Consumption, Production and Material Culture*. London: Bloomsbury.

de Solier, I. (2018). Tasting the digital: New food media. In K. LeBesco & P. Naccarato (Eds.), *The Bloomsbury Handbook of Food and Popular Culture* (pp. 54–65). London: Bloomsbury.

Domo. (2019). *Data Never Sleeps* 7.0. Retrieved from https://www.domo.com/learn/data-never-sleeps-7

Goodman, M.K., Johnston, J., & Cairns, K. (2017). Food, media and space: The mediated biopolitics of eating. *Geoforum*, 84: 161–168.

Lewis, T. (2018). Digital food: From paddock to platform. *Communication Research Practice*, 4(3): 212–228.

Lupton, D. (1996). *Food, the Body and the Self*. London: Sage.

Lupton, D. (2016). Foreword: Lively devices, lively data and lively leisure studies. *Leisure Studies*, 35(6): 709–711.

Lupton, D. (2017). Personal data practices in the age of lively data. In J. Daniels, K. Gregory, & T. McMillan Cottom (Eds.), *Digital Sociologies* (pp. 339–354). Bristol: Policy Press.

Lupton, D. (2018a). Cooking, eating, uploading: Digital food cultures. In K. LeBesco & P. Naccarato (Eds.), *The Handbook of Food and Popular Culture* (pp. 66–79). London: Bloomsbury.

Lupton, D. (2018b). 'I just want it to be done, done, done!' Food tracking apps, affects, and agential capacities. *Multimodal Technologies and Interaction*, 2(2). Retrieved from http://www.mdpi.com/2414-4088/2/2/29/htm

Lupton, D. (2019). Vitalities and visceralities: Alternative body/food politics in new digital media. In M. Phillipov & K. Kirkwood (Eds.), *Alternative Food Politics: From the Margins to the Mainstream* (pp. 151–168). London: Routledge.

Lupton, D. (2020). Carnivalesque food videos: Excess, gender and affect on YouTube. In D. Lupton & Z. Feldman (Eds.), *Digital Food Cultures* (pp. 35–49). London: Routledge.

Lupton, D. & Feldman, Z. (Eds.). (2020). *Digital Food Cultures*. London: Routledge.

Rousseau, S. (2012). *Food and Social Media: You Are What You Tweet*. Lanham, Maryland: Rowman Altamira.

Schneider, T., Eli, K., Dolan, C., & Ulijaszek, S. (Eds.). (2018). *Digital Food Activism*. London: Routledge.

Taylor, N. & Keating, M. (2018). Contemporary food imagery: Food porn and other visual trends. *Communication Research and Practice*, 4(3): 307–323.

Thrift, N. (2008). *Non-Representational Theory: Space, Politics, Affect*. London: Routledge.

Vannini, P. (2015). Non-representational research methodologies: An introduction. In P. Vannini (Ed.), *Non-Representational Methodologies: Re-Envisioning Research* (pp. 1–18). New York: Routledge.

INDEX

Page numbers in *Italics* refers figures; **bold** refers tables and followed by 'n' refers notes number